REVISED
Nuffield Physics
TEACHERS' GUIDE
YEAR 5

ISBN 0 582 04685 8

General Editors
Eric M. Rogers
E. J. Wenham

Contributors
D. W. Harding
J. L. Lewis
A. W. Trotter

REVISED
NUFFIELD PHYSICS
TEACHERS' GUIDE
YEAR 5

Published for the
Nuffield-Chelsea Curriculum Trust
by Longman Group UK Ltd

Longman Group UK Limited
*Longman House, Burnt Mill, Harlow, Essex, CM20 2JE, England
and Associated Companies throughout the World.*

First published 1967
Revised edition 1980
Second impression 1988
Copyright © The Nuffield-Chelsea Curriculum
Trust 1980
ISBN 0 582 04685 8

Design and art direction by
Ivan and Robin Dodd
Illustrations drawn by Hayward & Martin Ltd.

Filmset in Monophoto Plantin 110 by
Keyspools Limited, Golborne, Lancs.
Produced by Longman Group (FE) Ltd
Printed in Hong Kong

CONTENTS

Foreword

During the period in which this book was being prepared for publication, the Nuffield-Chelsea Curriculum Trust has been founded. This new Trust has been set up jointly by the Nuffield Foundation and Chelsea College, University of London, to continue the work of curriculum development in the sciences and in mathematics which the Foundation began and has maintained over a period of nearly two decades and with which the Centre for Science Education at Chelsea College has been closely associated for much of that time. It was the general success of the teaching schemes initiated by the Nuffield Foundation that has enabled the setting up of the Trust and thereby ensured that the work of curriculum change and development should proceed on a permanent basis, responding when necessary to the evolving needs of teachers and schools.

This book and its accompanying *Pupils' Text* complete the revision of the O-level physics materials. The other books, already published, are the *Pupils' Texts* for Years 1 and 2, for Year 3 and for Year 4, together with *Teachers' Guides* in parallel and an accompanying *General Introduction* to the course. The new Trust owes a great debt to the authors and those associated with the project in many fields. The co-operation of the Schools Certificate Examination Boards (in the present case the Oxford and Cambridge Board) has been an essential element in making sure that the aims of the course are fully recognized and tested in the examinations they provide. Among the many organizations who have given us help I must particularly mention the Association for Science Education, whose members, through their writing, conversation, and contributions in other ways, have brought to our attention the needs of the practising teacher and of the pupils in schools.

This new edition of the Nuffield physics material draws heavily on the work of the editors and authors of the first edition published in 1966. An immense debt is owed to them. The physics programme was inaugurated in May 1962 under the leadership of Donald McGill. It suffered a severe setback with his tragic death on 22 March 1963 but those who were appointed to continue the work have done so in the spirit in which he initiated it, and in the direction he foreshadowed. He was succeeded as organizer by Professor E. M. Rogers. Together with the associate organizers, John Lewis at Malvern and E. J. Wenham at Worcester, the assistant organizer, D. W. Harding, and the deviser of the *Question Books*, the late H. F. Boulind, the teams of teachers led by Eric Rogers produced teaching ideas that have influenced profoundly curriculum discussions and physics at a time of major educational change.

The new volumes draw in many ways on the original *Teachers' Guides* and *Guides in Experiments* and *Question Books*. Their contribution in providing a firm basis for these further developments is gladly acknowledged here. It is also a pleasure to acknowledge the part played by the large number of teachers who have helped in discussion and feedback and in making useful suggestions, but it is to Eric Rogers and Ted Wenham, who as General Editors have completed this work, that we especially record our thanks.

Lastly I should like to acknowledge the work of William Anderson, our Publications Manager, and his colleagues, and of course our publishers, the Longman Group Ltd, for their continued assistance in the publication of these books. The editorial and publishing contribution to the work of the projects is not only highly valued but central to effective curriculum development.

K. W. Keohane
Chairman,
The Nuffield-Chelsea Curriculum Trust

General Editors' Preface

Nearly twenty years ago the Nuffield Foundation, following requests from teachers who suggested changes in O-Level physics teaching, gave a large grant for studies of needs, development of apparatus and the provision of printed materials to offer a new teaching programme to schools who liked to try it.

The essence of that programme, as it emerged from consultations, visits to schools, discussions in groups of teachers, was a change from teaching hampered by insistence on rote learning towards even more learning for understanding which, it was felt, would provide greater chances of pupils' learning of science being transferred towards long-lasting benefits.

By now, pupils of many schools have tried that programme – we believe with enjoyment and some success. As pupils reached the end of the five years to face an O-Level Examination, the teaching proved justified by the admirably relevant Nuffield Physics papers produced by the Oxford and Cambridge Schools Examination Board (acting on behalf of all Boards). The number of candidates for that Nuffield O-Level Physics Examination is now well over 20,000 each year.

Those Nuffield papers were set with the aim of testing the teaching and learning that we suggested; and they received sympathetic marking which looked for understanding in candidates' answers.*

Many teachers have followed some general suggestions:

1 Let pupils work in the lab in small groups, often pairs, and leave them alone to make their own mistakes and find their own solutions, except where rescue is needed. That seems to us near to professional science.

2 Use stimulating questions as principal learning aids to encourage discussion, reasoning and imagination.

In making the revision for this new edition we received a general directive from the Foundation: that we should try to maintain the same standard of enquiry, and learning of science for understanding, and not change the programme in a way that would 'lose the Nuffield spirit'. The Foundation recognized the changes in school structure but considered that other programmes, such as Nuffield Secondary Science, provide better for other levels of treatment than a heavily diluted version of our programme could do.

We started the revision by consulting some 200 teachers, some of them in person, many by profuse enquiry forms. We also visited a considerable number of schools to see Nuffield classes in their present form. Again, those visits influenced us very profitably in our revision.

We changed Dr Henry Boulind's excellent Questions for thinking and understanding to simpler wording, but retained their essential enquiry. In response to pleas from teachers, and to the needs of the new school structure, we added Progress Questions to provide a different and easier approach.

Our most important change of all in the revision has been the production of the *Pupils' Text* in four volumes, to provide young scientists with help for experiments

---

* Two small examples may illustrate this.

i The Board prints on the front of the examination paper all the formulae likely to be wanted – this is an assurance to both teachers and pupils that just 'memorizing formulae' is not so important. Candidates realize that memorizing definitions and formulae is not very profitable. On the other hand, the Examiners expect a candidate to understand the origin and uses of some formulae and their limitations – like a capable craftsman. And they expect a candidate to be able to describe physical quantities and relationships in his or her own words.

ii In marking scripts for O-Level, the Nuffield Examiners have not felt themselves restricted by a fixed marking scheme. They read with a flexible attitude, looking for good knowledge, imagination, and interesting suggestions too – which they award with bonus marks.

and some discussions of ideas, also thinking questions and progress questions. Thus for many pupils this book should act as a complete substitute for work cards.

On behalf of teachers and pupils who will use these books, we owe thanks to many people: to our consultant teachers, without whose advice we could not have envisaged the needs of the project; to Professor R. A. Becher, who was our chief inspiration and guide in the original project, to whom we still turn for wise advice; to Professor K. W. Keohane as our co-ordinator with counsel concerning Physics and teaching and people; to John Maddox, till recently Director of the Foundation, for past interest and care, and now special encouragement. Both teachers and pupils will owe much to the five teachers who constructed the Progress Questions – forged and tempered them: Anthea Arnold, Margaret Fawcett, Reinet Fremlin, Gwen Jones, and Hilda Misselbrook.

We would also wish to thank Ann Sinclair and Truda Temkin for their sterling contributions; our editors, Elizabeth Bland and Hendrina Ellis, who saw the books through the stages of production; Mrs A. D. Powell and her team at Hayward & Martin Ltd, for the illustrations; the designers and art directors Ivan and Robin Dodd; and Deborah Williams who carried out the picture research.

Our thanks are also due to Paul Black, John Harris and Bill Trotter for their constructively critical advice and to Dr Peder Moesgaard of Aarhus University for his helpful review of the Astronomy Section.

All who have contributed hope that this new form of the programme will enable many of the next generation to enjoy physics and to remember it all their lives.

Eric M. Rogers, E. J. Wenham
General Editors

Preface to Year 5

Plans and Hopes

Before describing the structure of this Year, let us take stock of our position.

Examinations ahead In Year 5 many pupils will be preparing for a public examination and this will inevitably influence the work to some extent. *But we hope that examinations will not dominate the teaching.* For a large fraction of our pupils this will be the last year of formal instruction in science. Some of those dropping the subject will be leaving school: others will go on to pursue non-scientific disciplines. What sort of scientific background do these people need?

A final year for some Consider first those who will be leaving school at the end of this year. Apart from those who enter engineering apprenticeships, few will make direct use of their scientific knowledge, so facility in experimental techniques is not of high priority; but they should know what it is like to do an experiment – which most of them enjoy – and appreciate something of the difficulties of interpreting the results. The ability to solve numerical problems is not a skill which is likely to survive the passage of years (dearly though one would like to think that the future householder could make a quantitative assessment of, say, the relative cost of different ways of heating his home). But it is of first rate importance to realize that physics *is* a quantitative science, in which it is possible to forecast correctly, by computations with known data, what would happen in a hypothetical situation.

Educated citizens As a citizen in a scientific world, a pupil should neither be afraid of science nor be overawed by it. He should learn that natural phenomena usually have a rational explanation and that scientific methods can be powerful tools in understanding and controlling man's environment. In other words, he should have an educated person's understanding of science. This is the end to which our scheme is directed. The choice of topics to be included is relatively unimportant – we show representative samples of physics, not the whole field. And we should remember that at the age of fifteen, most pupils will grasp concrete examples more readily than abstract principles.

Essential knowledge In addition, there are certain key ideas that are so important, both in physics and in the World as a whole, that they should become second nature to everyone. These are:

The conservation of energy and the dominant role that energy plays in scientific theory and in the economy.

Heat as a mode of molecular motion, and the statistical nature of thermal laws.

The properties of electric currents (conceived as a flow of charge in a stream of electrons).

The nature of light and the properties of the whole range of electromagnetic radiation.

The atomic nature of matter; the existence of fundamental particles.

The nature of radioactivity and nuclear changes: both the possibilities and the dangers of these.

The growth of atomic theory, from early pictures to modern models.

Anyone with this background should be able to listen to a scientist talking in general terms and follow at least the gist of his argument – perhaps even pick his brains. That is an ability which should make life more interesting and meaningful for the average citizen and we hope that some pupils, at least, will be impelled to find out more for themselves.

In addition, many people – business managers, lawyers, shop stewards, nurses – may have to carry on a discussion with technologists in the course of their careers; and parents and teachers will have to answer the questions of young people. A scientific background is, if not essential, at least highly desirable in every walk of life.

Some will continue There is nothing in the discussion above that does not apply equally to the future arts student and the future scientist – good science for citizens is also good science for specialists – but the latter require something more. Nowadays, the student of almost any of the humanities needs some knowledge of science. The historian must be aware of the impact of scientific knowledge on the thought and economy of the period he studies. The archaeologist uses scientific tools in his work. The economist is concerned with science as an economic force – also he aims at using scientific tools and analogies. The philosopher is increasingly concerned with scientific matters. To cater for these it is important to include something of the history of science, especially as new methods of teaching may obscure the original approach.

It is probable, too, that the future arts student will appreciate a panoramic view of modern science (though the potential science specialist might be content with traditional treatment), and he will certainly want to look closely at the way in which physical arguments are justified and how they hang together. We hope that these things will be taken up again, at a greater level of maturity, but the basis should be laid now and the relevance of science to humanities made plain.

These aspects too will be valuable as part of the general education of school leavers, although some of the topics are more academic and theoretical than would be chosen for that group alone.

If one topic in this Year seems too hard to a pupil, other topics are likely to appear too hard also. If some pupils find the early topics too hard, we recommend an immediate change of balance towards more experimenting.

Yet, if a *teacher* considers the topics too hard for his pupils at a first glance, we hope that as an experimental scientist he will try teaching it *twice*; a first round to see its possibilities, a second round, the following year, to see how his own version runs.*

* This showed up strongly in the early trials of the first edition where astronomy was taught directly by the teacher. Some teachers who were at first uneasy about the astronomy came to delight in it and greatly enjoyed teaching it in a later round.

Future science specialists Last, we must make some provision for the needs of future physicists, chemists, mathematicians, engineers, doctors, *et al*. This is a very important group, since the economic well-being of the country will depend strongly on their professional ability. It could be argued that the whole syllabus should be designed for specialists, and the others left to make what they can of it. This has been tried far and wide. It has often made physics unpopular, and that is not our policy. We believe that, up to O-Level, education should be general and not vocational, and that the needs of ordinary people, as citizens and individuals, should predominate – even over our strong economic demands.

Nevertheless, it is important that the programme should lay a broad and firm foundation upon which later specialization may be built, and that nothing be done to turn the potential scientist from his path. We hope our physics may attract some of the waverers into a scientific career.

We trust that the future scientist will be interested above all in the *ideas* of science: he will be a poor scientist if he is not. It must be recognized, however, that many will be more attracted by the power over the environment that science places in their hands. So our teaching should include topics catering for them and giving some facility in experimental techniques.

Formulae And the proper place of formulae – as servants, not as masters – should be taught. The employers of school leavers entering engineering apprenticeships will expect such preparation. For the very bright pupils, we need to include some quite difficult problems to cultivate their intellectual interests, to show that science is worthy of their mettle.

A broad syllabus Thus our syllabus must cater for many needs. Inevitably, some compromise is necessary. It is hoped that examinations will allow the teacher to emphasize the aspects which are best suited to the interests of his class. Some aspects will be treated mainly in laboratory work, or in pupils' reading, or in homework problems, where the emphasis can be changed to suit the class.

Applications We have made no explicit mention of topics in applied science. This programme does not base its teaching on them directly. Yet pupils should be aware of the way in which physics interacts with engineering, and we should show them something of the nature of the latter – one of the prime needs of the country is to make sure that young people do not despise applied science. We therefore recommend that the topics of our syllabus should be illustrated, wherever possible, by examples of their application. We believe – not from laziness, but from our survey of needs – that the examples should be chosen by each teacher according to his interests and those of his pupils, rather than from examples listed in a manual.

A Very Long Guide

This *Guide* is long and discursive. That is intentional, because these notes are offered to many different teachers with varied interests and experience, for guidance in following a new programme of teaching.

Where one teacher wants to know our reason for suggesting a topic, another may want to know why we advocate some crude apparatus instead of a modern machine; and elsewhere, why we recommend a strange modern machine instead of simpler traditional apparatus.

Some teachers may welcome detailed instructions for running an experiment. Others will be distressed by the lengthy discussions of details, and they will ask for a short list of topics, such as the following:

Motion in a circle: acceleration directed to the centre

Measurement of e/m for electrons

Planetary astronomy and gravitational theory

S.H.M. (qualitative only); waves, alternating currents

Interference of light: Young's fringes

Diffraction grating; spectra; electromagnetic spectrum

Radioactivity – properties of radiations using electroscope and counters

Alpha particle scattering and Rutherford atom model

Photoelectric effect

Theories of light: waves and photons

Matter waves: particle and wave behaviour

Newer atomic models

Given thus, in a few lines, our list can hardly satisfy a teacher who is planning a new programme with changes of aims and attitudes – in examinations as well as in teaching. At most it tells an external critic our topics, without telling him our intentions!

Because our suggested programme will be new to some readers and the format for treatment of this final year will be unfamiliar, we shall enter, at some point, into long discussions and give considerable details. We trust that some teachers who would prefer a quicker summary will bear with us and will extract from that profusion whatever they need.

The Work of this Year

This is a Year of important experiments and ideas, in which we draw upon the work of previous Years but expect more imaginative thinking, more reasoning, and new experimenting. We want to develop some taste for theory and to explore further in atomic physics in both experiment and theory.

Newton's Laws and orbits Newton's Laws of Motion – so far treated as general principles and tested in simple class experiments – are now put to the use that Newton himself set forth: to form a grand theory of the Solar system.* For that a *quantitative* treatment of circular motion and its central acceleration is essential. (Instead, the result, $a = v^2/R$ could be announced – simply asserted – then tested experimentally, but that would prove tedious.)

Newton's great work deserves serious study. That is one reason why we offer a whole section of *Pupils' Text* on astronomy. Another reason is that we need to show pupils a physical theory being built up and used. (We have to teach about atomic theory with too much assertion; that is why we prefer astronomy for our prime example.)

The history of planetary astronomy is a special topic for the class to explore by reading, each pupil at his own speed and level. Yet it deals with one of the greatest intellectual developments in the

---

* Newton did not evolve his Laws to solve Newton's Law problems, such as calculating the force when a car hits a tree! He stated them as basic assumptions, starting points in building his grand conceptual scheme of astronomy: grand because it was so economical and fruitful – essentially only four basic assumptions, then a vast outflow of explanations and predictions: Kepler's Laws, the shape of the Earth, explanation of tides, the Moon's motion. . . .

scientific world. As A. N. Whitehead put it after writing on just this topic:

> The moral of the tale is the power of reason, its decisive influence on the life of humanity. The great conquerors, from Alexander to Caesar, and from Caesar to Napoleon, influenced profoundly the lives of subsequent generations. But the total effect of this influence shrinks to insignificance if compared to the entire transformation of human habits and human mentality produced by the long line of men of thought from Thales to the present day, men individually powerless, but ultimately the rulers of the world.*

Electrons Then, armed with some understanding of orbital motion, we can continue previous work on electron streams by bending their path with a magnetic field. To analyse measurements, pupils must use some knowledge of the force exerted by a magnetic field on a stream of charged particles. That is difficult, but we shall not evade it (thus losing our chance of clear knowledge of electrons), or spoil it by announcing an unexpected 'formula'.

Instead we shall make a direct experimental approach and measure the strength of the magnetic field that we use by putting a simple current-balance in it.

The experiment leads to a comparison of electrons with atoms. And it will remain as useful background to which we can refer when we mention similar measurements: for ions in a mass spectrograph, for alpha particles and beta particles from radioactive sources, etc.

Waves and oscillations A study of waves and oscillations is resumed from earlier Years. Pupils see, and measure, interference patterns. This is intended to do three things:

(i) Give factual knowledge of waves and interference, which is an important part of one's general knowledge (and it is a beginning for some A-Level physics).

(ii) Let pupils see for themselves why we think light consists of waves, and let them make their own estimate of the wavelength of light.

(iii) Provide a necessary background for introducing a topic of quite modern physics: matter waves.

Science and the modern world by Alfred North Whitehead, Cambridge University Press, 1926, pp. 299–300.

If we expect to have time later to mention this phenomenon and discuss it briefly and gently, we must have prepared pupils beforehand by making them familiar with the behaviour of waves with gratings.

Pupils take a short qualitative, experimental look at simple harmonic motion. This helps, on one hand their knowledge of wave behaviour – for use in building an atom model – and on the other hand their work in the lab with alternating currents – for use in ordinary life. They continue from earlier Years' experiments with transformers, power line, and slow a.c.

Teachers may feel tempted to go straight on from discussions of interference and gratings to theories of light and the contrasting behaviour of quanta or photons. That lies ahead, and we hope pupils will hear some of it, because it is an essential part of our modern view. Yet, before we proceed to that, we have another topic, which will contribute.

Atom models and radioactivity While atomic models are being developed, experiments on radioactivity will be carried out. This work will open up new knowledge and help to encourage the imaginative thinking by which scientists formulate a 'model'.

We continue our building of atomic models, from the stage of the hard, round elastic molecules or atoms that sufficed in kinetic theory to a picture of Rutherford's nuclear atom.

Should we go further still? As well as giving pupils some understanding of energy and some proficiency with electricity do we not owe them – particularly the many who will finish their learning of physics now – some more modern knowledge of the microphysical world? We could describe the atom model which has emerged from the ideas and experiments of photons, matter waves. . . .

That may seem too difficult; yet here is a *fable* to give a warning and encouragement.

A fable We have tried to make our course include some of the physics of today. Rather than emphasize the atomic physics of half a century ago, we suggest bringing the teaching nearer to the present day, even for O-Level pupils.

Imagine a conference on the teaching of physics, convened in AD 1700. A resolution might well be passed to the effect that the teaching of Aristotelian mechanics is in good order and should continue; teachers in schools are skilfully expounding the dynamic principle that 'motion requires a force proportional to velocity' and have good apparatus to demonstrate it. The new ideas of Newton should be relegated to advanced seminars in universities.

Now imagine a conference in the early 1800s: the teaching of Caloric would be endorsed and Rumford's unorthodox view of heat as connected with motion – with the new name 'energy' about to appear – would be viewed with suspicion and restricted to postgraduate discussions.

Shift our imaginary conference on teaching physics to the early 1900s. Newtonian dynamics, energy and its conservation, atoms, molecules and kinetic theory, are all being taught clearly and well; but measurements of electron streams are regarded as very difficult to teach, and the rumours of a quantum restriction are pushed away to professional studies. Special relativity seems a misfortune.

The lag is natural enough: in each generation the older material seems to be secure knowledge and easy to teach well; and the newest material is not only strange, but as yet, difficult to teach because its teaching vocabulary has not been developed. Of course that is partly due to the different way in which teachers have learnt it. In many cases the older material was taught them in their own student days with firm authority – and, if they were given some of that material at a sufficiently early age by a strong capable expounder, they may have accepted it uncritically. Whether we like it or not, we must accept that as one general characteristic of education – we who are teaching now must, without knowing it, be giving a strong dogmatic force to some of the physics we are teaching.

On the other hand material that a teacher did not learn in student days is apt to remain a little strange and not seem so strong a part of the syllabus.

For example, many an older physicist today regards Relativity as somewhat uncomfortable – however well he now understands it and perhaps teaches it. When he first met the new ideas of Relativity they struck him as regrettable; well assured geometry was being attacked and could be shown to be 'wrong'. But to the new generation of physicists, Relativity is a commonplace, heard about at school, used as a normal part of student physics.

Thus, the lag is there and forgivable, and in past ages it may have been harmless. There has been time for each generation to catch up. Now with science growing and changing so rapidly, and ideas travelling so fast round the world, is it any longer safe to let teaching lag in this comfortable way? Trying to make the teaching catch up and lessen the lag would be uncomfortable and even dangerous, if done carelessly. Yet when we move our imaginary Conference on Teaching Physics to the year 2000 we may feel uneasy about the prospect. Will so much of today's newest physics still seem too strange to teach?

With that question in mind we urge teachers to continue courageously and to give their present pupils glimpses of photons, matter waves, and a wave-determined atom model, perhaps even a mention of uncertainty. We give some sketches and short descriptions in *Pupils' Text*.

Teachers may wish to have fuller accounts of such topics. Yet when they look at books on modern physics they are disappointed. There are up-to-date advanced texts for university teaching or professional use; and there are some popular accounts of the latest physics, written for laymen. Many a book which gives the careful exposition of modern physics that one would like to have as background for O-Level teaching seems to stop short at the state of physics sixty years ago, or at least treats later topics too briefly. So we list suggestions for books in Appendix 1 at the end of this *Guide*.

Experiments in this Year's Programme

Links with earlier years The following are essential now if they have not been done fully in previous years.

Millikan's experiment – discussion and film. This should be done in two parts: first, a clear proof that electric charges come in multiples of a single universal basic charge; second, a measurement of the size of that charge. The first part is both more important for our present teaching, and easier to show, though even that will have to be shown by film. The measurement of the value of e will have to be taught by assertion; details are probably outside the normal scope of the programme.

Interference of waves: Young's fringes by ripple tank (class experiment); Young's fringes for light (qualitative class experiment); Young's fringes for light, rough measurement (class experiment).

Electron streams: straight line in a vacuum; deflection by *electric* fields.

It will not be necessary to do experiments to illustrate the relations among FORCE, MASS, and ACCELERATION, even if pupils missed them. However, pupils must not only know $F = ma$ and $Ft =$ change of (mv), but must also have an understanding of the nature of mass, force, weight, gravitational field strength, and kinetic energy. They must know that K.E. $= \frac{1}{2}mv^2$.

Pupils need not do or see experiments with electric fields even if they missed them, provided that they know the pattern of the field between parallel plates and are ready to accept the idea that field strength E is given by p.d./(distance between the plates).

Experiments this Year The teaching during this Year involves some important experiments: measuring a model satellite orbit; a test of $F = mv^2/R$; measuring the wavelength of light; measuring e/m for electrons; and some radio-activity experiments. These should be class experiments as far as possible. Some class experiments with a.c. and transformers have been postponed from Year 4 till now so that pupils can enjoy working at them carefully: experiments with the electromagnetic kit; and experiments with slow a.c. – all with plenty of use of oscilloscopes. There are quick qualitative experiments on S.H.M.

Even so, those will not occupy the full amount of time deservedly allowed for laboratory work. We suggest as a major laboratory activity, that pupils should be allowed to choose and give 'demonstrations for revision'. Teachers will find this plan outlined in a note at the end of Chapter 2. We offer a list of suggested experiments, from all the Years of the programme, in Appendix 2 at the end of this *Guide*. Average pupils (heading for O-Level) will gain much in setting up and running a careful choice of demonstrations.

Some, with special interest and ability, may wish to choose a 'grand' experiment, such as Millikan's experiment done on their own, or a measurement of the speed of light – one such adventure can make a tremendous contribution to a young person's education. It need not make great demands on the teacher's time once the apparatus is provided – in fact, it *should not* do so, since the point of the exercise is not to train the pupil in advanced experimenting but to give him the experience of independent work. Teachers with heavy timetables and crowded laboratories may think this an unrealistic dream; but we believe that some pupils who have followed our programme in spirit as well as in content will be ready to undertake such work in a trustworthy and re-sourceful way that will make that dream come true.

A Mixed-Ability Group

By this stage the physics class in many a school will have changed its proportions; it will have become partially selective by pupils' own choice. Although in Years 4 and 5 most schools will continue a policy of having classes with a wide range of abilities and interests, pupils will have had a choice of pro-grammes and only some will have chosen to continue physics. Each has been guided in choosing (usually, though not always) by his interests and sense of his own abilities.

Sorting by choice will often come at the beginning of Year 4, so we may expect at least three-quarters of the present physics class to be promising Nuffield O-Level candidates. The remaining quarter will be kept there by some other pressures.

There will be some who may not do well in a final examination. For them carefully chosen pupil *demonstrations* done all on their own can provide confidence, enjoyment, and occupation, to take the place of work that seems hard or uninteresting. Never mind whether their difficulties are from lack of interest or poor ability, they deserve special care. They are *in* Nuffield physics and they deserve an equal opportunity to *enjoy* physics. The chance that they will benefit and transfer some gains to later living must depend on some measure of enjoyment. If they find the learning and reasoning boring and puzzling – too hard – it is our duty, for the sake of good education, to offer them easier, creative activities.

Examinations and Revision

Revision will, of course, be a problem for each teacher to judge in terms of his class and their work, also in terms of Nuffield examination questions and their marking. We do not suggest that the Year's work should go right up to the examination without revision, just because the atomic physics at the end is so important. Yet we do believe that many teachers will find, when they get to Year 5, that the Nuffield examination questions devised to fit our programme do not need the same type of revision as the traditional ones.

True, the examiners will dip back into the work of earlier Years for questions; but in doing so they will look for understanding in the sense discussed in the *General Introduction* that was published with *Teachers' Guide 3*. There we gave a general account of our aims in teaching for understanding, and suggestions for moving towards those aims: let pupils learn by doing their own experiments, let them argue things out (with help), and let them solve problems that ask for thinking.

We suggested that taking more time for one topic to gain a sense of mastery, might give more chance of lasting understanding than shorter glimpses of several topics.

However, such general descriptions of teaching are not very helpful when we are thinking about the actual examinations. The comparison that was offered in terms of the French verbs, *savoir, connaître, comprendre*, was at best a helpful admonition. But we also gave a relevant and useful definition: we reminded readers that most of us say, at one time or another, 'I never really understood that part of physics until I came to teach it', and we suggested that, in the same sense but on a much simpler scale, the test of a pupil's understanding of a topic can be whether he can teach it.

We elicit his teaching by asking him to explain something to someone else – his younger brother or his non-scientist uncle, rather than to a mysterious, stern examiner who requires the knowledge to take on a formal polish.

We have been using that device for questions all through our programme, both for current teaching and as preparation for questions like those in examinations. If, as we trust, O-Level examinations for our programme continue to be slanted in the direction of asking the candidate to teach things in his answer to a benevolent enquirer, they will have a good chance of testing understanding.

Of course, such questions have always been used by good examiners but our suggestion here is that the questions should take a less formal style and that the answers should be read by examiners with this requirement of understanding still more actively in mind.

In marking the answers, examiners will find they need to read flexibly. Sometimes they will have to make subjective judgements, since they are looking for the understanding that the pupil shows in the answer – and for a general feeling of mastery – rather than for memory of facts. Thus examiners will be doing great good, on behalf of our teaching in particular and science education in general. They may find that marking schemes of very precise form are unsuitable for some questions; but they will be able to judge whether the pupil *understands* – in much the same way as many of us judge in an interview whether an applicant understands the work he is to do.

With such hopes in mind, we urge teachers to encourage creative thinking and good experimenting by pupils, and to move away from a great deal of revision of factual material which might not be so useful in examinations.★ As with many things in our suggested programme, this is a matter where the first round of teaching may be difficult and uncertain, and teachers will find that they know far better what to do in a second round.

Specimen examination questions We reprint in this *Guide* a selection of Nuffield O-Level examination questions, by kind permission of the Examining Board.† In studying those questions, teachers should remember that some questions intentionally ask about an experiment or phenomenon that pupils have not met, in order to test intelligent thinking. Teachers should tell pupils beforehand that such a question only asks for simple thinking.

Pupils' demonstrations for revision We suggest that a good deal of pupils' time in the lab (say, one-third of the year) should be devoted to their setting-up of experiments that they have seen or heard of before, making their experiment go, then showing it to other pupils who come visiting. Pupils should work in pairs, or quartets at most. They should emerge from their work ready to explain clearly; but they should not make written records (except for any measurements and calculations) because that would take too much time and prevent them from doing more demonstrations.

★ One critic has commented: 'In preparing for Nuffield examinations, traditional last-minute or last-month revision is little more than a *placebo* – for pupils and for the teacher.'
† Teachers can obtain copies of both Papers I and II for a small fee from: The Oxford and Cambridge Schools Examination Board, 10 Trumpington Street, Cambridge, CB2 1QB.

The public reason for this activity is temporary and clear: revision. Our private reason is a long-term one: personal experience enjoyed and to be long remembered.

For further comments, see the note at the end of Chapter 2. For suggestions of experiments, see Appendix 2 at the end of this *Guide*.

Too Full a Programme?

Reading our outline, teachers will wonder whether this is too full a programme. We consider it is feasible because there are two aspects that combine to save time and help progress: the arrangement of ASTRONOMY AS PUPILS' READING for homework and the use of PUPILS' DEMONSTRATIONS FOR REVISION.

Early in the autumn, pupils should start reading the astronomy section. A teacher who follows this reading himself will find opportunities for giving encouragement and help with occasional demonstrations.

While astronomy occupies homework or some spare time in class, pupils should choose experiments to set up for themselves and then show them to the rest of the class. These pupil-demonstrations for revision may well continue through the Year, but they should start early. They will provide much more relevant revision *for our kind of examination questions* than the usual formal revision.

If Year 3 has prepared for Year 4, and Year 4 has had its full time, pupils and teachers *will* cover Year 5 comfortably.

Comments on the Structure of this Year's Programme

This is a year of putting physics to work to build stronger knowledge, principally in understanding of theory and in atomic or nuclear physics.

We do not intend to provide new topics compactly taught for examinations or to spend a major part of the time revising old topics. However, we shall survey a good deal of new atomic physics, and pupils' reading of astronomy will be novel material for most. In earlier Years pupils will have learnt many regions of physics on which examinations can draw, with questions that ask for constructive thinking. This Year should give pupils practice in such thinking at a more mature level.

Thus the attitude should be: 'Now we can extend and use earlier knowledge to tackle great problems concerning the structure of the world.'

New tools and new knowledge Essentially this programme introduces six new tools and uses them together to develop five areas of physics:

1. We discuss motion in a circle and arrive at $a = v^2/R$ and $F = mv^2/R$.

2. We obtain from experiment a quantitative measure of the force exerted by a given magnetic field on a current in a wire, and we extend that, by argument, to a charged particle moving across a magnetic field.

3. The idea of an inverse-square law is introduced for gravitational fields but is applicable with the same geometry to electric fields, the spreading of light, etc.

4. Devices using ions are used to exhibit 'atomic' events: cloud chambers, Geiger counters, spark counters, etc.

5. Alpha particles from radioactive substances (or protons or electrons from accelerators) are used as projectiles with which to explore atomic structure more deeply.

6. Studies of water ripples and light are combined to provide new criteria for waves.

With these tools, we develop:
a. Quantitative knowledge of electrons, positive ions, and nuclei as parts of atoms.
b. An example of physical theory – seen in stages of construction. Pupils read the history of man's knowledge of the stars, Sun, Moon and planets from early observations through successive stages of gaining knowledge and building 'theories', to the age of Kepler and Galileo, when man had a great body of empirical information, organized in rules that were verified with precision but were still disconnected pieces of knowledge. Then pupils meet the unrolling of Newton's great gravitational theory to see the fruitful use of good theory in science.
c. Knowledge of radioactivity: contributions to atomic models; uses and hazards.
d. The wave:particle idea. We touch briefly on the modern picture of both radiation and matter having particle aspects and wave aspects – the behaviour which we observe and measure being determined by our choice of experiment.

We cannot pursue this duality far with pupils at this age, but we should introduce our modern view, both for the sake of non-scientists who will read about such things later and to set the stage for further studies by physics specialists.

e. Atom models: we develop successive models from hard billiard balls adequate for simple kinetic theory to a hollow Rutherford model; then a survey of later developments.

We owe modern knowledge to our pupils, but the experiments and reasoning that led to such knowledge (even if we show simplified forms) are too complex for present teaching. We can only offer a survey of results and descriptions of some models. However, in this region of modern developments we feel justified in breaking our resolution to insist on providing supporting experiments and we suggest giving only short descriptions.*

Our General Aim

All through, the important thing for teachers to keep in mind is the overall view that they are giving to pupils who will finish with physics now; the knowledge of physics that each young 'scientist for a day' is gaining, with a picture of nature explored and well understood up to a point, then bounded by new regions of unfinished knowledge. Here at the end, as in the earlier Years, we hope pupils will conclude that 'science makes sense'.

Satellites: Chapters 1 and 2

We use the mechanics of motion round a circular orbit for several topics in the course of this Year: Earth satellites, measurement of e/m for electrons, a description of mass spectrometers, a reference to measurements on alpha particles and beta rays; as well as its essential part in Newtonian astronomy.

Therefore it seems essential to show pupils the derivation of $a = v^2/r$. That is given in Chapter 1. We offer a second version in the form of a structured question and we apply the result to an Earth satellite.

As a preliminary, ask pupils to draw a satellite's orbit to scale and find its period simply by treating the motion as a case of free fall. With the orbit close to the Earth pupils give the acceleration the ordinary value of local g; so they need not know or use $a = v^2/r$.

After the derivation and a discussion of 'centrifugal' vs 'centripetal' pupils can deal with satellites further out.

In Chapter 2 pupils participate in a measurement of e/m for electrons in a fine-beam tube; and they compare the result with e/M for hydrogen ions in electrolysis – then we suggest that an electron is only a small chip from an atom. Millikan's experiment, which we trust has been shown by film in Year 4, adds further knowledge.

As a preliminary to the main measurement we remind pupils of the 'catapult force' on a current in a transverse magnetic field, then we modify that description to apply to a charged particle moving across a magnetic field.

A note about pupils' demonstrations for revision is placed at the end of Chapter 2, to precede the Astronomy section.

Since the Astronomy section is intended for pupils' reading, we suggest it offers a good time for pupils to start demonstrations for revision in the lab.

The plan is described in that note in this *Guide*, also in a note in *Pupil's Text*. And a list of suggested experiments is given in the Appendix at the end of this *Guide*.

*** Films?** We might give occasional support or elucidation by films, but there are two dangers:
(*i*) A film which shows the real apparatus and working of a fundamental experiment may be merely confusing, owing to the profusion of auxiliary equipment.
(*ii*) A film which describes either ideas or experiments by *animation* may be very misleading in another direction. It can sketch the story we *think* or *hope* is true and fail to give real teaching of science. However tempting such a film looks as a clarifier, it would be unwise to show it.

CHAPTER 1
Motion in an orbit

INTRODUCTION TO CIRCULAR MOTION

Experiments and questions Start with three demonstrations and a class experiment.

1. Show an example of constant velocity. Ask whether this is 'natural motion' and whether any forces are needed to keep it going.

Demonstration 1
CO_2 puck sliding with constant velocity

Apparatus

1 magnetic ring puck	item 95C	
1 glass plate	95A	
4 wedges	95B	
1 CO_2 cylinder	19/1	
1 dry ice attachment	19/2	

Preparation

Clean the glass plate carefully with methylated spirit or window cleaning fluid, and polish it with a duster. Level it carefully with wedges. Test the levelling with a puck.

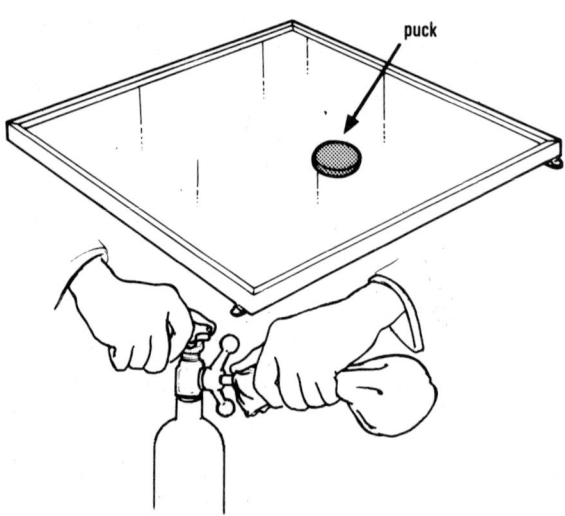
puck

Procedure

Make a small quantity of dry ice using the CO_2 cylinder* and dry ice attachment. Put it under the puck on the glass plate. Give the puck a gentle push. Pupils watch it travelling from end to end –

and back again to show that the motion is not maintained by a sloping table!

2. Refer to the motion of the Moon round the Earth. Ask whether it is 'natural motion' and whether any force is needed to keep it going; leave the question unanswered.

3. Then feeling the way towards the general need for a force for orbits, let each pupil whirl a small block of wood or rubber on a string and decide about the force – is it a push or a pull on the block?

Experiment 2
Whirling a small satellite

Apparatus

32 small blocks of wood†
32 pieces of string, about $\frac{1}{2}$ metre long

† The centripetal force kit includes rubber bungs, which will be used soon. But small blocks of wood make simpler objects for this first trial; having one per pupil will make better teaching than having one rubber bung per pair.

Procedure

Each pupil whirls his 'satellite' above his head in a horizontal circle. If he pulls the string in a little, he may feel its tension more easily.

At some stage, *Pupils' Text* reminds pupils that 'strings pull, they never push'.

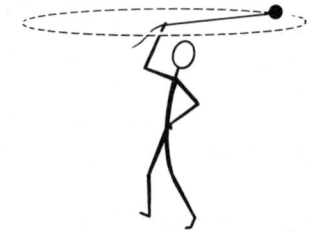

* The special form supplied for our experiments has a central pipe inside, like a soda-water siphon: keep it the right way up and release *liquid* CO_2.

Examples Ask pupils which way the force acts on them if they sit on a smooth seat in a car that rounds a sharp corner. Which way do they slide? Which side of the car then pushes on them?

Perhaps discuss the banking of a bicycle rounding a corner, but this often leads to more confusion than help, because the problem is better discussed by taking moments than by considering a single force.

'Flying off' Now or later, give a very important demonstration and discussion: the motion of an object released from its orbit.

Demonstration 3
Stopping the motion: releasing a satellite

Apparatus

1 wooden block or rubber bung
    string, 1 metre

Procedure

Tie a piece of string about 1 metre long to the wooden block. Swing it round in a horizontal circle of radius about $\frac{1}{2}$ metre.

Explain that it will be released at an instant when the block is nearest to the class. (It is useful to say, like this: '... now ... NOW ... NOW', and let go on the third 'now'.)

Release it suddenly at the right instant. It will travel tangentially, to the side of the room, not radially out at the class as some pupils expect. Some may flinch because they are sure centrifugal force will drive the block out towards them. Point out, now or later, that the block is not driven out by centrifugal force. Nor does it *fly off* along the tangent in an aggressive way; it just *continues* its motion steadily along the tangent in a straight line as soon as it is left alone – an example of Newton's First Law.

Centrifugal force? Centripetal force? Teachers will find themselves immersed at once in questions of centri*fugal* force versus centri*petal* force.* Be sympathetic and firm: ask the direction of the force on a stone being whirled in a circle. Insist once more that 'strings pull, never push'.

Experiment 2 (repeated)
Whirling a small satellite
Let pupils try this again, to illustrate the discussion.

Ask again which side of the car pushes on the passenger on the smooth seat. Even ask, rather unfairly, 'What do you think the Earth does to the Moon, repel it or attract it?'.

Explain that the same string which pulls the stone *inward* also pulls the holding hand *outward*; and that the sliding passenger in the car will push the outer side of the car. And we agree that the Moon must then pull the Earth outward – think of tides.

Some critics will maintain that there is therefore a *centrifugal* force acting on the stone, the passenger, and the Moon. And we ourselves would certainly infer the presence of such an outward force *if we were in a rotating frame of reference*, riding on the stone, or in the car.

In the case of a rider on the Moon, where Earth's gravity is the controlling force for the orbit, the evidence for a centrifugal force would be suppressed, because the observer himself would be pulled by the Earth with a force proportional to his mass, so he would have the same orbital motion as the rest of the Moon.

We should have to say that centrifugal force is one way of looking at the problem; but not our way.*

---

* As pronounced with the customary stress, the two names centrif'g'l and centrip't'l sound too much alike. As a temporary trick, teachers may find it safer to mispronounce them, by agreement with the class, centriFUGUE'l and centriPET'l.
* The choice of policy between rival treatments, centrifugal and centripetal, has advocates on both sides.

In advanced physics, we all call on centrifugal force. We reduce a problem of orbital motion to a statement of equilibrium by adding an outward, centrifugal force to the forces applied by strings, gravitation, etc. (D'Alembert force.) And when we explain the action of a centrifuge (or a merry-go-round) in detail we all want to resort to centrifugal force, though of course a logical centripetal explanation can be given.

For an elementary beginning, some teachers prefer to use centrifugal force because it draws on pupils' common belief. Others hold that this start will lead to difficulties. They maintain that, having started with the view that acceleration a

Special note on centrifugal force

In elementary teaching we must make a clear decision between two approaches to motion in an orbit: the appeal to centri*petal* force and the idea of centri*fugal* force. A mixture of both is fatally muddling for beginners.

Centripetal force used with Newton's Second Law will yield the right answers, and forces will always be in the right direction – for example, strings will pull and never push: lorries rounding a corner will skid or fall outwards. But the method may seem artificial to pupils, who have all heard of centrifugal force. The following discussion with an imaginary pupil may be helpful to teachers dealing with this question.

Motion round a circle needs a real inward force, provided by real external agents. This view of centri*petal* force will help you to deal with all real problems of motion in an orbit.

Then what is centri*fugal* force? You often hear of it, you may find yourself speaking of it when you whirl something round, and you will find some books using it to explain things in physics. Here are several opinions on it. Choose according to your taste.

Opinion I: *Centrifugal force is a phoney force, imagined through a misinterpretation of evidence, a confusion between agent and victim.*

If you whirl a stone on a string, the string's tension pulls your hand outwards (just as it pulls the stone inwards). This is a real centrifugal force *on your stationary hand*, not *on the whirling stone*. You feel your hand being pulled outwards, so you say, 'I feel the stone and string pulling my hand outwards. That tells me that the stone is being pulled outwards, by some centri*fugal* force, and the string is just transmitting that force.' That is where you are mistaken. There is no outward force on the stone. Really the string, in a state of tension, pulls at both its ends. While it pulls your hand outwards it pulls the stone inwards. The only real force *on the stone* is inward, centri*petal*.

needs force F in the same direction ($F = ma$), we should continue to take that view for the acceleration v^2/R, which is certainly centripetal, not centrifugal.

One thing is sure: in elementary teaching a mixture of the two approaches is fatally confusing.

In planning the course, *the Nuffield Physics Group decided to set forth the suggested teaching in terms of centripetal force. A separate note in a box discusses the choice at length.*

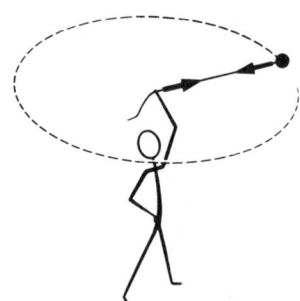

Preparation for OPINION II

Suppose that two boys, A and B, visit one of those amusements at a fair in which people sit on a floor that rotates. A and B enter the room while the floor is at rest, and sit on the polished floor. Knowing the trick of the performance, A glues himself to the floor. When the floor begins to spin A notes that a mysterious force seems to pull him outward. But for the glue, it would make him slide out to the wall.

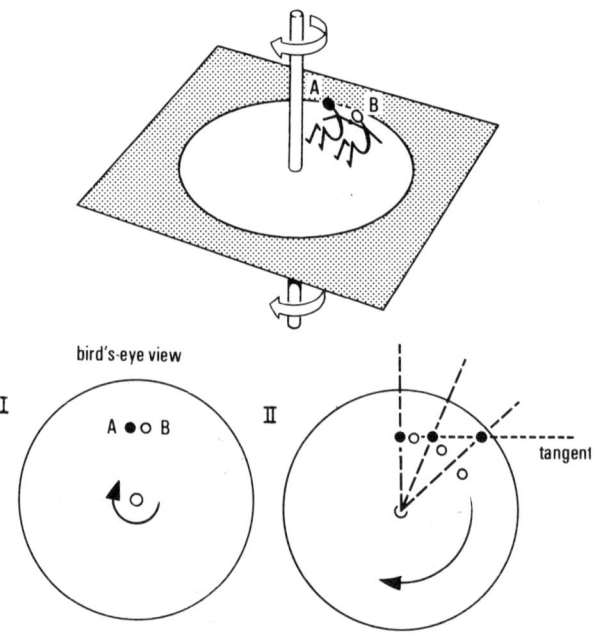

B, without glue, slides out to the wall if A does not hold on to him, exerting an inward pull on him. Each boy feels he is struggling against 'centrifugal force'.

But now let a stationary observer take a bird's eye view from above. Seen from outside the spinning room, A and B are each moving in a circular orbit, and each needs a real *inward* force to keep him in orbit. For B, the force is the inward pull which A provides: for A it is the pull of the sticky floor on him. The boy A merely imagines an outward force on B because he has to apply a real inward force to him. As the outsider sees, these inward forces are not neutralizing a mysterious outward force, they are *making an inward acceleration; they are making A and B move in a curve.*

The outside observer offers a further comment. He sees that when A lets go, B continues along a tangent (if there is no friction). B's successive positions along that tangent are farther and farther out from the centre of the circle; so, as seen by A (revolving with the floor), B *seems* to be sliding out along a radius. But really B is just *continuing a straight (tangent) path, a simple example of Newton's First Law.*

Opinion II: *Centrifugal force is a delusion due to living in the rotating system and trying to forget it.*

The rotating-floor discussion leads straight to this view. To people sitting on the floor in a concealing fog – and ignoring its motion – there is an outward field of force, endowing every mass m with an outward force mv^2/R. Unless some real agent applies an inward force to balance this, any object left alone will seem to slide outward with acceleration v^2/R. We prefer to take a sober view from outside and say that both the outward field of force and the outward sliding are delusions due to *living in a rotating framework and not allowing for its motion.*

Opinion III: *The Novice's Headache Cure*

Here is a good use for centrifugal force. Let us be rude and say, with some truth, that some beginners prefer 'statics', the physics of things at rest (in equilibrium), to the physics of motion. Problems involving acceleration and rotation make a novice's headache. He wishes that such problems could be reduced to the simple statics problems that he is so good at – forces in bridges and cranes for example – and they can. Consider, for example, the problem of a pendulum whirling around in a conical motion. The two real forces acting on the bob are its weight and the string tension. These two real forces must add up to a resultant force mv^2/R inward – otherwise the bob would not revolve round the orbit. Here then are two forces W and T which have horizontal resultant mv^2/R inward. Now turn this into a statics problem with resultant zero by adding an extra fictitious force. *What fictitious force must we add to W and T to*

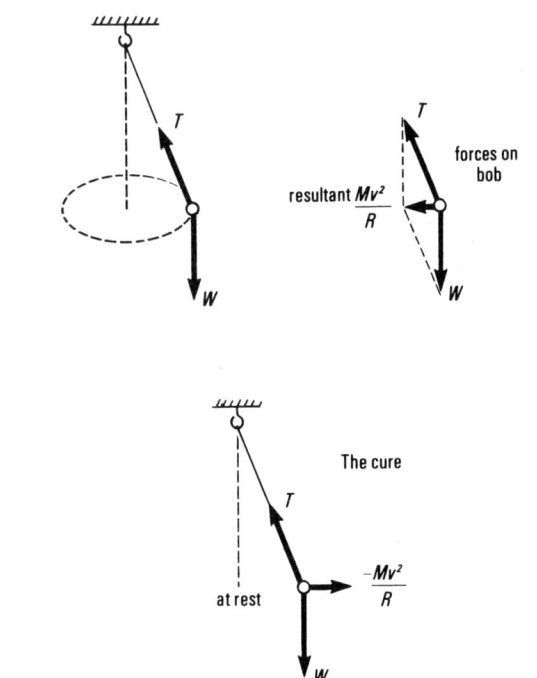

make zero? The third force would have to be $-mv^2/R$, meaning outward. So we might say this to the novice: 'Yes, you can turn any problem with circular motion into a statics problem if you *take all the real forces acting on the moving body, ADD a fictitious centrifugal force, mv^2/R outward, and then write an equation stating that these forces (including the fictitious one) have resultant zero.* Solving the equation will give you the same information as the method of making the real forces combine to produce inward acceleration v^2/R'.

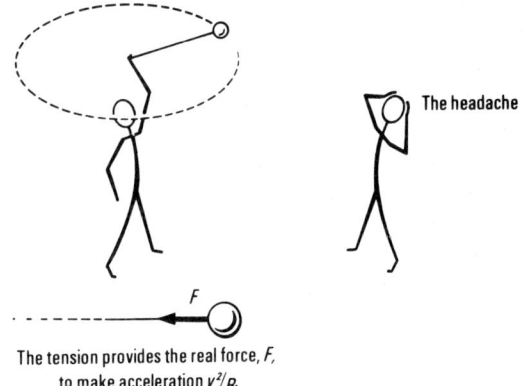

The tension provides the real force, *F*, to make acceleration v^2/R.

In this treatment, centrifugal force is a fictitious force, but a useful one to cure the novice's headache.

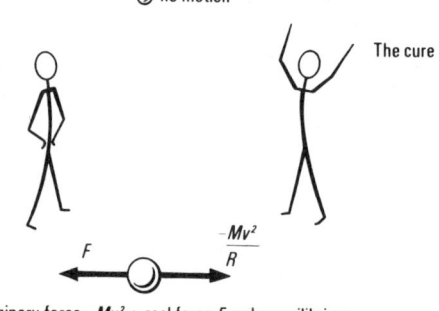

no motion

The cure

F $\dfrac{-Mv^2}{R}$

Imaginary force $\dfrac{-Mv^2}{R}$ + real force F make equilibrium

Centrifugal force is also used thus in advanced physics, to save trouble – but then it is a sophisticated trick in the hands of skilled craftsmen. As used by most students, it gives the right answer but makes some of the theory harder to understand – how can that be avoided when it reduces obvious motion to fictitious equilibrium? The trusting user, with his right answer, is confused about the forces: he is not sure which are real or which way they pull. *If you value your understanding of physics, avoid this headache cure at all costs.* Of course, a *mixture* of this centrifugal headache cure with centripetal forces will produce utter confusion!

Opinion IV: *Relativity*

(This opinion sketches some comments from sophisticated Relativity theory. Read it for amusement or for a good moral warning, but do not let it convert you to the headache-cure method for novices. This view is true, but only within the framework of definitions constructed for it.)

Can nothing better be said of centrifugal force? Returning to Opinion II, some scientists ask, 'Why is it so wicked to view things from a rotating framework? After all, we live on a spinning Earth. Are the 'centrifugal forces' that arise from our rotating-framework viewpoint really different from

other forces, and less real? Who are we to say which is really rotating, ourselves or everything else?' (We are back to Copernicus *vs* Ptolemy, with Einstein to offer a view.) This last question is like the problem of testing Newton's laws in an accelerating railway train. Simple experiments in mechanics would give strange results: a pendulum would hang at a tilt; the surface of water in a pail would take a similar slant; and experiments with dry ice or a trolley on a horizontal table would not show Newton's Laws in simple form – an unexpected constant force would be added. Yet all those peculiarities would disappear if we built a suitably tilted room inside the train for our lab. Then we could still find the usual laws, though we should find 'gravity' changed in size and direction. We suspect that we *cannot* distinguish between the effect of acceleration and a real change of local gravity – Einstein built General Relativity theory on an elaboration of that 'cannot'.

Relativity theory starts with an axiomatic statement, that we cannot tell which is moving, ourselves or 'the other fellow'. And that makes us modify the simple geometry of space and motion that Euclid assumed and Galileo and Newton used. For constant velocities, there are Relativity tests which fail to distinguish absolute motion, even with the help of signals by flashes of light. So we feel justified in accepting the Relativity principle and its modified geometry. In practical life, the modifications are not perceptible, and they only affect experiments noticeably when very high speeds are involved, as they are in atomic physics and perhaps in astronomy.

Extending the Relativity attitude to accelerated motion we assume that a local observer will find the effects of acceleration indistinguishable from a local change of gravity; thus we decide that gravitational fields can be treated as local changes of geometry in space-time. This is Einstein's

constant acceleration

Principle of Equivalence. Though the viewpoint is new, its practical form shows only small deviations from Newton's law of gravitation.

Extending this idea to rotation, we suggest that a local observer cannot distinguish between the effects of a rotating framework and a local change of gravity, if he is moving with that frame. In that case centrifugal force tugging outward would be just as real to him on the spinning floor as an extra, horizontal pull of gravity.

Then, to a tiny creature in a centrifuge, centrifugal force fields would appear just like real gravitational fields, only some thousands of times as strong as ordinary gravity. And gravity would take on a new direction – he would quite forget about its old direction. This General Relativity view has proved useful in co-ordinating thinking; and so far we have not observed anything inconsistent with it. In such a way, centrifugal force has grown to be respectable, and when we want to test the effects of large gravitational fields, unattainable on Earth, we think we may use a centrifuge instead.

The general Principle of Equivalence forbids us to call the rotation of the Earth absolute. It therefore leads to a new mechanics that will predict the same effects whether the Earth spins and moves round the Sun, or the stars and Sun move round us. On General Relativity theory, a rotating universe would produce 'centrifugal forces' at a stationary Earth; so tests of a spinning Earth, with a Foucault pendulum or equatorial changes of 'g', could not distinguish between the two causes: spinning-Earth or everything-else-spinning. Faced with the old question, 'Is Copernicus right and Ptolemy wrong?' we must demur at Galileo's cocksure insistence and say, 'Both views *may* be equally true, though one is a simpler description for practical thinking and working.'

Opinion on the Four Opinions?

Make your own choice. However, for problems and your present work, you are advised to use only centri*petal* force.

We shall use centripetal force We insist that we shall treat the problems ahead of us by the clumsy, unrealistic-looking method of saying that anything moving in a circle must be acted on by an inward, centripetal force that pulls it in from a straight-line path. A real *inward* pull is needed – 'No force, no orbit'. Before this has time to build up irritation or boredom, proceed to a satellite.

SATELLITE

To start our study, we might say to pupils:

Throw a cricket ball out horizontally. It falls to the ground some metres away. A rifle bullet, fired faster but also horizontally, reaches the ground after a kilometre or so.

Try a thought experiment: fire a bullet so fast that it covers still more of the Earth's circumference before it reaches the ground. What effect has the Earth's curvature on the bullet's fate? Fire it fast enough to make it fall over the edge of the Earth, so that the Earth falls away from the bullet's original direction just as fast as the bullet does.

Now, to the bullet, all parts of the Earth are the same, and it soon forgets where it started from. Given just the right speed such a bullet will always be 'falling over the edge', and it will go on and on round the Earth – keeping just above the ground – until it arrives back at the starting point and hits us from behind.

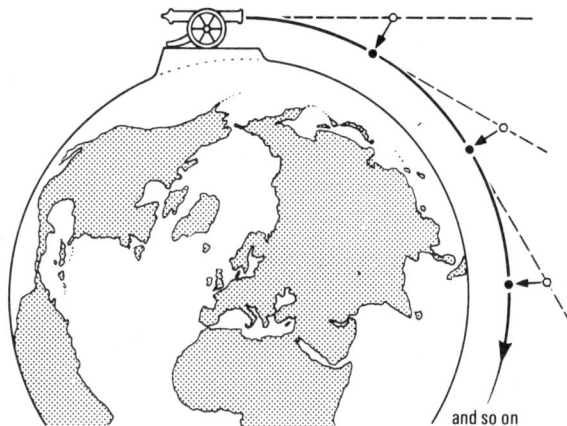

and so on

In practice, air resistance absorbs energy, and down comes the bullet. So a successful satellite must start orbiting outside the atmosphere.

Here is a simplified story for a typical satellite that we may give a class that is interested:

1. Rocket starts off nearly vertically. The exhaust gases exert an upward push greater than the rocket's weight, so the rocket accelerates upward.

2. Fuel exhausted, motor cuts out, and the first stage separates.

3. Parabolic (free fall) trajectory until the path is horizontal at maximum altitude.

4. Final stage ignites and accelerates its relatively small mass to high velocity before unlatching the satellite proper and leaving it in orbit. (Exhausted final stage is also in orbit, but, in the course of time, a small relative velocity puts a big distance between them.)

5. There is some air resistance even at 200 km up, so energy is 'used up' slowly and the satellite descends. In the course of this descent, its time to circle the Earth grows smaller. Pupils may try letting a string carrying a whirling mass wind up round a finger. They will see it speed up.

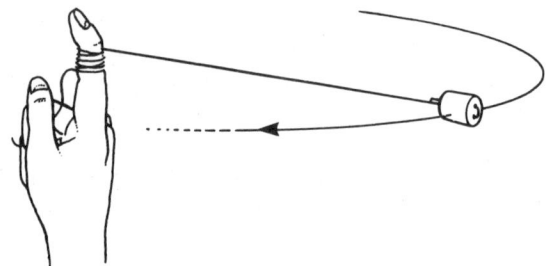

Questions about satellites and rockets will flourish now, whether we want them or not. We should welcome them and perhaps use some of them to lead to the topics ahead.

A matter for discussion Ask about the energy necessary to raise a satellite through 200 km. Would it be better to burn fuel slowly, giving thrust for a long time, or burn it rapidly and then have a long rise time under free fall conditions? The answer lies in the cost of raising the load of fuel in the first stage. If fuel-burning continues during most of the rise, we have to raise a good deal of fuel. We have to compromise between saving that cost by a rapid initial acceleration and the stresses on man and machine involved by large acceleration. At the final stage we must provide energy $\frac{1}{2}mv^2$ for the motion in orbit. All the latter – in fact, all the energy released by the fuel – is dissipated as heat on re-entry.

Centrifugal force again Even now the question of centrifugal force will crop up with strong advocates. Teachers may like to try two more attacks on it:

1. Ask pupils to think about a boy running along a straight path with a larger, stronger boy running

beside him and pushing him steadily sideways. What would be the effect of that continuous sideways push? Suppose, as the smaller boy changes his path, the larger one continues to push sideways in a direction perpendicular to the new path. What kind of path would the smaller boy take? A circle seems reasonable. In this case the force is clearly inward.

Point out that there is always a visible (or, if invisible, well-known) agent applying a force towards the centre, such as a string that pulls inward, or the pull of gravity on a satellite.

2. Loop the loop. Pupils should see a demonstration of looping the loop, not just as a piece of circus entertainment but as a basis for valuable discussion at this stage. *Centrifugal* force will be strong in the popular vocabulary, but this experiment can help to convert to *centripetal*.

Demonstration 4
Looping the loop

Apparatus

1 flexible curtain rail†	item 119
1 steel ball	131A
many retort stands and bosses†	504–505

† Excellent ready-made loop-the-loop rails of plastic appear in toy shops from time to time, with names such as 'hot wheels'. These are preferable because the supports are simpler and less obtrusive. (However, we do *not* advise schools to buy special exhibition models of heavy metal channel.)

Preparation (of home-made model)

Bend the curtain rail so that the ball can loop the loop after being released high up on the rail. This can be made with a rail 2 metres long, but it is more effective with a 3 or 4 metre length. The initial descent should be very steep and the loop needs to be tight.

It is worth while to construct a mounted version with the top section of the loop detachable. Mounting is not easy. A convenient method is to glue blocks of wood 4 cm × 4 cm × 4 cm at 30-cm intervals along the rail. Glue the blocks with Araldite or similar adhesive, and reinforce with screws (counter-sunk so that the steel ball does not hit the screw-head as it runs on the rail). The wooden blocks can be drilled with holes to take rods or nails attached to retort stands with bosses.

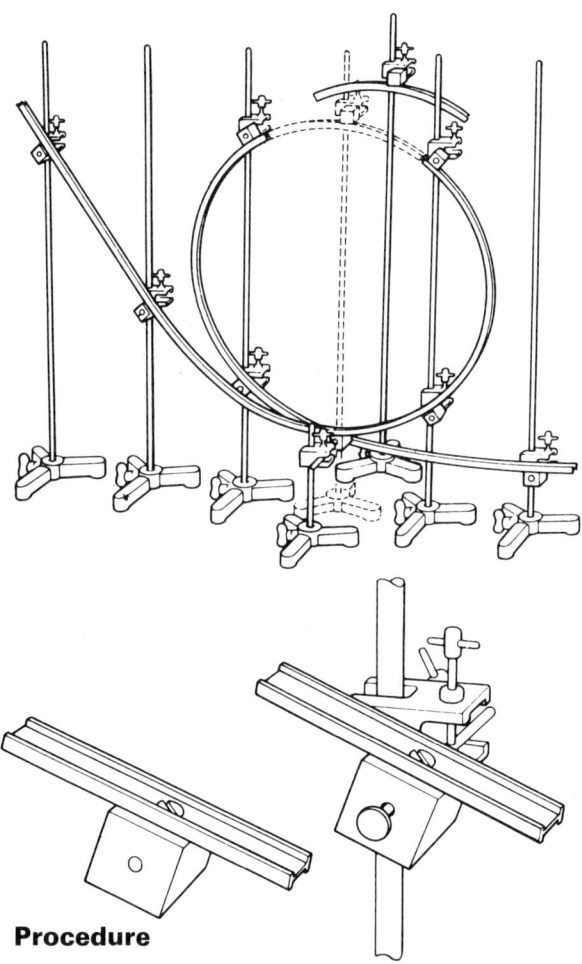

Procedure

(*i*) *The demonstration*: release the ball high up on the launching section.

The noise of the ball running on the track is a welcome part of the performance – all the better if the ball makes its final landing silently. Arrange the track with a short straight section after the loop, tilted slightly upward, and place a tray of

sand to catch the ball when it leaves the track.
(*ii*) Release the ball still higher up; more noise, equally successful looping.

Discussion

Hold the ball on the rail at A on one side of the loop and ask, as in the *Pupils' Text*:

★   ★   ★   ★   ★

What compels the ball to go round a circle? What pushes or pulls it with a real force to make it do that? What direction has that force, and what provides it? It must be some real inward force, towards the centre of the loop.

What provides that force at A, half-way up the loop? It must be the rails that push inward. Does gravity also act on the ball when it is there? Yes, of course. But gravity pulls vertically, and its only effect is to give the ball a downward acceleration so that the ball slows a bit.

What provides the inward force at B, the top of the loop? The rails may push downwards, but what other force helps? Yes, gravity helps fully now whether it is wanted or not!

Look what happens when we have the ball moving much slower, needing less force for its orbit. Gravity provides too much force and makes the ball fall away from the rails.

★   ★   ★   ★   ★

(*iii*) Show the motion with various speeds, and find the critical speed at which the ball *just* follows the loop, with the rails exerting no force at the top of the loop because gravity suffices.
(*iv*) Ask what would happen if the rail were cut out just at the top of the loop. If possible, remove the top section of rail and show what happens.

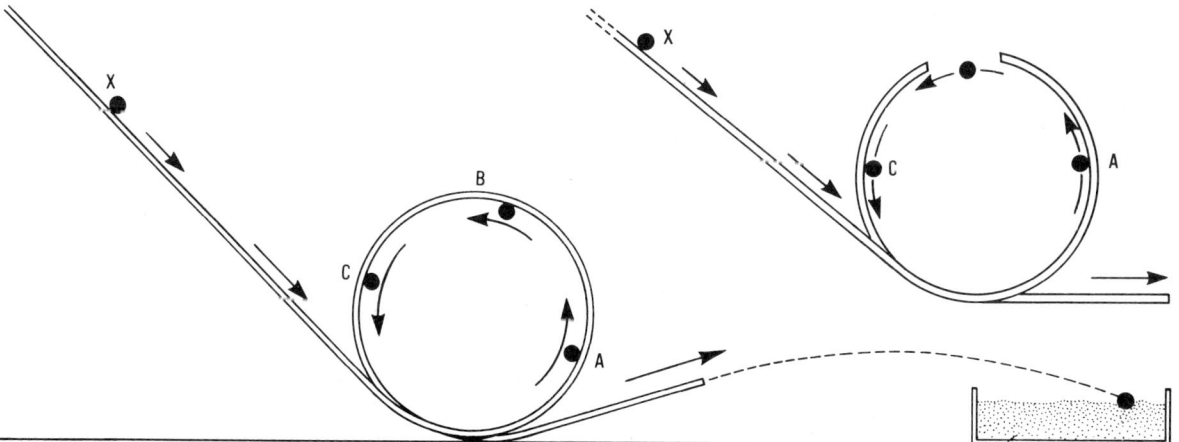

An important experiment for pupils Let pupils draw a satellite orbit to scale, treating the motion as a case of free fall with ordinary g. Thus they estimate a satellite's period without using $a = v^2/R$ yet.

Class Experiment 5
Sketching a satellite orbit and predicting its period

Apparatus

16 sheets (+ many spares) of brown paper or lining
    paper,† each about $1\frac{1}{2}$ metres long, 15 to 20 cm wide
1 length, $3\frac{1}{2}$ metres, strong, thin wire
1 large 'weight' or hook to anchor one end of the wire
    on the floor or near it
16 metre rules with cm and mm markings

† As sold in wallpaper shops.

Pupils should work in pairs or alone. Working in larger groups will court failure. It may be best to work on the floor.

Procedure

Pupils draw part of a circular orbit for a satellite close to the Earth, 200 km up. They assume the satellite is always falling inward with the ordinary acceleration g (≈ 10 m/s per second), so they do not need to know or use $a = v^2/R$ at this stage.

Taking $a = 10$ m/s per second and $R = 6400 + 200$ km, pupils calculate (with help, if necessary) the satellite's free fall from the tangent to its orbit in a chosen time of 120 seconds, called x km. [Our private answer is 72 km.] We recom-

mend a scale of $\frac{1}{2}$ mm to 1 km for the drawing – a smaller drawing would be difficult to measure.

With that scale the radius of the orbit is $6600 \times \frac{1}{2}$ mm, or 3·3 m. Pupils draw an arc about $1\frac{1}{4}$ m long with that radius, or have it drawn for them. A thin wire anchored on the floor at one end and held taut with a pencil at the other end serves for this. (A 3 m length of string would stretch enough to make the drawing harder. Even so, a 7 cm error in the radius will make less than 1% error in the final result.)

That arc is twice as long as pupils need for 2 minutes' travel, but the double length enables them to make a symmetrical drawing★ which will yield a good estimate more easily.

On that arc, XAY on the sketch, pupils mark the mid point A and part of the radius to A. They drop down along that radius their own calculated fall x km (to scale), AM. [Our private value $\frac{1}{2} \times 72 = 36$ mm.] They draw the tangent at A, symmetry helping. And they draw a chord XMY parallel to the tangent, with mid point at M.

Then they can transfer the fall AM out to the place where it should be shown as a fall from the tangent to the orbit, a fall NY. This determines the point Y so pupils can measure the travel distance AY, which is covered in their chosen time 120 seconds. [Our private answer: about 973 km.]

They know the total travel distance for the whole orbit, $2\pi R$, which is $2\pi 6600$ metres; so they

★ Careful drawing or calculation shows that such a satellite travels almost 1000 km in 2 minutes. So a double length of arc, for symmetry, needs to be twice $100 \times \frac{1}{2}$ mm, or 1 m long; but pupils' estimates and drawing will vary considerably, so the arc should be drawn at least $1\frac{1}{4}$ m long. There is of course no need for the centre of the circle to be on the paper.

It is unwise to change to a less simple scale in order to fit a smaller piece of paper. The arithmetic will bring enough troubles without that.

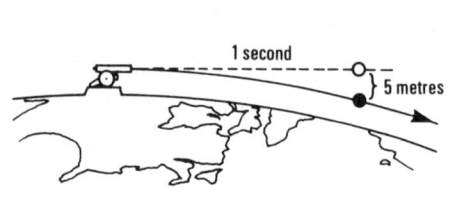

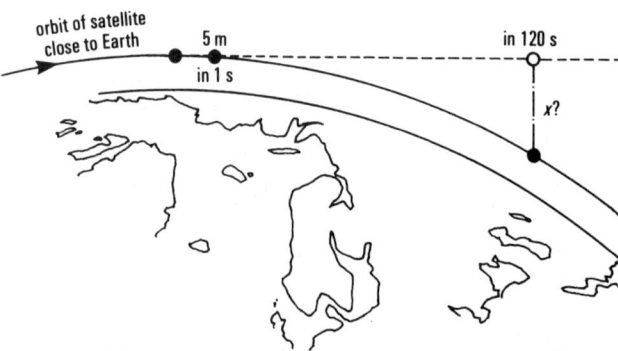

can now calculate the time that will take. [Our private answer about 85 minutes.]

The experiment sounds complicated, but pupils will find detailed instructions in *Pupils' Text*. If a pupil likes to have a second try, after struggling with the drawing and calculation once, he will find it quick and satisfying.

Does gravity extend to the Moon? Some teachers may wish to encourage sophisticated enthusiasts in the class to extend the satellite test to the Moon, first without, then with, inverse-square gravity. The same 3·3 metre arc will serve for a piece of the Moon's orbit, if we change the scale. (This is not described in *Pupils' Text*.)

Experiment 5X
The Moon's orbit time
(*BUFFER OPTION*)

Tell pupils that the Moon's distance from the Earth is about **60** Earth-radii. Then, for the

Moon's orbit, instead of a scale of $\frac{1}{2}$ mm to a kilometre, we have $\frac{1}{2}$ mm to **60** km.

First, try imagining that gravity extends undiminished out to the Moon. Calculate the fall in 2 minutes: 72 km as before. On the new drawing with $\frac{1}{2}$ mm to **60** km, that fall would be only $\frac{72}{120}$ mm – too small to work with. Therefore suggest letting the Moon travel the same arc on the diagram as for the Earth satellite falling for 2 minutes. (That was somewhere between 40 and 80 cm, according to each pupil's success in making the drawing.)

But now the fall of 36 mm from the tangent to that place *on the diagram* no longer represents 72 km. It represents a fall of 72 × **60** km. How much time would a falling body need for that, under full gravity? Look at $s = \frac{1}{2}at^2$. If we make s **60** times as big, with the same a, t^2 must be **60** times as big. Then t must be $\sqrt{60}$ times as big: that is, $7\frac{3}{4}$ times as big.★ Then the Moon would travel the arc in $7\frac{3}{4} \times 2$ minutes; and it would travel the whole circle in $7\frac{3}{4} \times$ the 90 minutes that pupils obtained for the satellite; that is, between 11 and 12 hours.

★ Since this is a difficult discussion, teachers who propose to try it with their class are advised to make sure that a rough value for $\sqrt{60}$ is known beforehand. $7\frac{3}{4}$ will serve well.
$(7\frac{3}{4})^2 = (\frac{31}{4})^2 = \frac{96\frac{1}{16}}{16}$ and $\frac{960}{16} = 60 \quad \therefore \sqrt{60} \approx 7\frac{3}{4}$ within 0·06%.

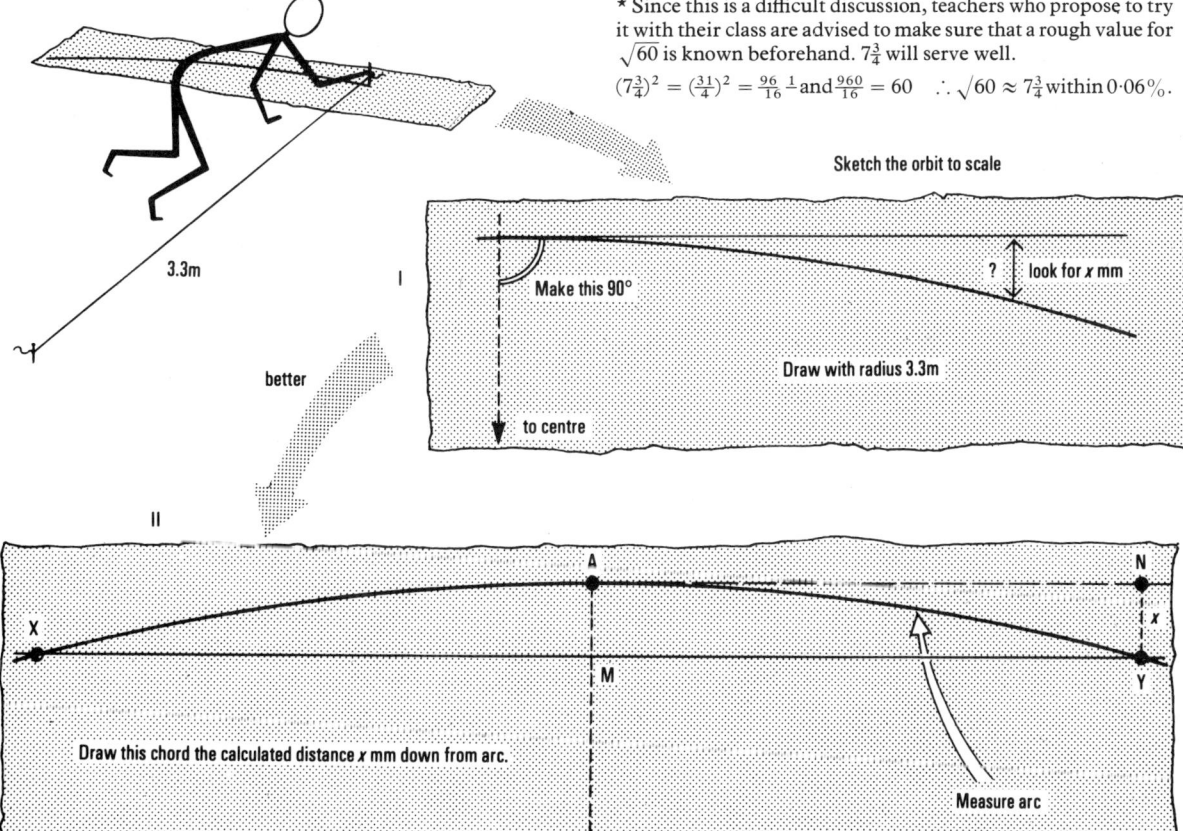

Sketch the orbit to scale

3.3m

better

I

Make this 90°

to centre

Draw with radius 3.3m

? look for *x* mm

II

X

M

A

N

x

Y

Draw this chord the calculated distance *x* mm down from arc.

Measure arc

Even if pupils' answers for the close satellite varied far from 90 minutes, the new answer is clearly wrong for the Moon, which takes a month. Therefore, if the Moon is constantly pulled from the tangent to its orbit by the Earth's gravity, *there must be a greatly diluted strength of gravity out at the Moon.*

Then pupils might take the Moon's month of 27·3 days for granted and calculate the amount of dilution. But the answer would not look clear and simple to beginners; so it is probably better to *suggest* an inverse-square dilution and try it in the calculation from the drawing, saying to the pupils:

★   ★   ★   ★   ★

If it *is* the pull of gravity that holds the Moon in its orbit, making it fall from the tangent to the orbit again and again, *it must be a much weaker gravity.* The acceleration must be much less than 10 m/s per second out at the Moon.

When astronomers started puzzling about this, several people suggested that gravity may 'thin out' according to an inverse-square law. According to that, if gravity is so much at a certain distance, it is $\frac{1}{4}$ at double distance, $\frac{1}{9}$ as much at treble distance, ... $\frac{1}{100}$ as much at 10 times as far away from the attracting body.

An apple near the Earth is pulled so strongly that it falls with acceleration 10 m/s per second. The Earth attracts an apple as if all the Earth were concentrated at the centre 6400 km below the surface, a whole Earth-radius from the apple.

But we know that the Moon is about **60** Earth-radii away from us, **60** times as far from the Earth's centre as an apple. So, if gravity does follow an inverse-square law, it must thin out by a factor $(\frac{1}{60})^2$ when we change from apple to Moon. If so, free fall under gravity at the Moon would not have an acceleration 10 m/s per second; but it would have acceleration 10/**60²** or 10/**3600** m/s per second.

How would that change affect your calculation for the Moon? Look at $s = \frac{1}{2}at^2$. If we make a **3600** times smaller, then, for the same s, t^2 must be **3600** times bigger: and t must be **60** times longer. As a result of diluting gravity, we should expect the Moon to take **60** *times our previous estimate for the whole orbit.*

That is **60** × (**7$\frac{3}{4}$** × 90) minutes,

or **60** × (**7$\frac{3}{4}$** × 90)/(60 × 24) days:

very close to 28 days.

It looks as if the Moon may be 'falling', to keep its orbit, with inverse-square-diluted gravity.

★   ★   ★   ★   ★

To arrive at that result, we must use the inverse-square law – otherwise all we can say is that *undiluted* gravity is much too strong.

We shall describe and discuss the inverse-square law and use it in our Newtonian prediction of Kepler's Third Law. However, to bring it in at this introductory stage may be discouraging; so we suggest that this extension of the orbit-drawing experiment to the Moon should be offered only to advanced pupils. We should tell them what the inverse-square law is, but we should not give a long lesson on it, and we certainly should not say that the inverse-square law is right, and that they ought to know it already. Instead, our pupils should follow Newton in using this discussion to see whether gravity *does* 'thin out' in that particular way.

How big is the force needed to maintain an orbit? We explain that to deal satisfactorily with satellites or electrons, we must know just how much force is needed to hold something in a circular orbit.

Common experience suggests: the higher the speed, v, the bigger the force needed to hold the objects in orbit, and the bigger the inward acceleration will be. And, for the same speed, the smaller the radius, (the sharper the curve), the bigger the force, and the bigger the acceleration must be.

So we expect the central acceleration to go up with v and go down with increasing radius. In fact $a = v^2/R$. We offer a geometrical proof of this for those pupils who can understand it easily. And we shall ask all pupils to give it an experimental test.

Necessity for mv^2/R If we restrict ourselves to a *qualitative* description of motion in a circle, and the forces it involves, we can talk generally about satellites but we cannot account for Kepler's Laws and we certainly cannot show Newton's great synthesis for the solar system in any clear light. We can describe what is shown by demonstrations with electron streams but we cannot make any measurements and so must stop short at a very general picture of atoms. Measurements of beta particles and the working of a mass spectrometer would remain vague.

So we must show pupils how to arrive at $a = v^2/R$ and thence $F = mv^2/R$ because we want that for several uses. If we proposed to show this simply as a piece of physics to be used for examinations, we should certainly find it difficult for many O-Level pupils – it would seem to them an odd piece of geometry rather than an essential piece of knowledge. As we use it here, it is essential knowledge, and we should face the difficulty of providing it.

We suggest that teachers should try the geometrical derivation of $a = v^2/R$ with any group that does not find it much too hard, starting by showing clearly why it is needed. Introduce the problem by showing how it is needed for satellites or for the electron stream in a magnetic field, then spend plenty of time on the geometry and algebra, so that the derivation becomes familiar by repetition. Then pupils should put it to such uses that they will consider the derivation worth while in retrospect.

It is easy enough to propose this for able groups, but what about the average group for whom the derivation will remain puzzling? Even for them, we suggest that this is something to try once, or at least to see done.

Few of us intend to climb Mount Everest, but we can all appreciate an account of the expedition and join in it by reading or by seeing a film, at least to the extent of understanding some of its hardships and enjoying some of the successes. If at the start, we remove the bogey of 'being examined'* and assure pupils that this is something they should see, (and even try doing for themselves), but not something that we propose to hammer into a compact shape that can be reproduced in examinations, our pupils should be old enough by now to appreciate this as valuable experience.

We hope that teachers will experiment with this approach – this method of talking about what one is going to teach, about the aim and method of its teaching, before embarking on the teaching itself.

In this, one is doing little more than following the good practice that any teacher adopts when he is explaining something to an adult. He does not try to drive home every stage of his story until his adult listener could reproduce it; nor, on the other hand, does he pare away the essentials of the story so that the adult says it makes no sense. Like such adult listeners, our pupils should be able to say, 'I have seen that. It was difficult but it was sensible and from now on I can take it on trust – trust vouched for by what I have seen myself.'

Then there are slower pupils, or some with less interest, for whom teachers feel convinced that any geometrical or algebraic derivation would prove much too puzzling. Even in such cases we hope that teachers will try the geometrical derivation to see whether, released from the examination bogey, the class can appreciate it after all. (We know no way of finding out whether that is feasible with a given class except by trying it.)

Where the derivation must be avoided, we have three choices: (1) to give up most of the physics of satellites and planetary systems, and atomic particles; (2) to treat those things, but take out all quantitative discussions, so that physics seems to lose its backbone; (3) to justify $F = mv^2/R$ by an experimental test. We hope that teachers will experiment with the last method. We can provide apparatus and suggestions for its use – though we wonder whether a slow pupil will not have as much difficulty in following the experimental test as in following a carefully taught derivation.

Derivation of v^2/R The teacher should preface this derivation by a considerable discussion of the general idea, and a reassuring statement about watching and seeing it done, so that science is not a mystery.

A sense of success We advise teachers to assure pupils that a Nuffield O-Level examination would not ask them to remember and reproduce a proof of $a = v^2/R$ in full – though there might be simpler questions on its meaning and use. A similar assurance is given in the *Pupils' Text*. Yet for some pupils we offer contrary advice: mastering this proof can give a keen pupil great satisfaction.

There are not many cases of a 'difficult proof' in our programme, which offer an opportunity like that. In Year 4, the discussion of random walk, mean free path, etc., leading to the size of a molecule is too hard to be suitable. The proof of K.E. $= \frac{1}{2}mv^2$ is an easier offer but probably less rewarding. This proof of $a = v^2/R$ with its demands of geometry, a little algebra, and some clear thinking seems intermediate – hence the following suggestions for helping keen pupils to learn it.

* Examiners in Nuffield O-Level Physics or the equivalent would expect pupils to show that they have had sensible contact with this topic; but they would not ask them to reproduce a complete proof from memory.

Teaching the proof would be best done in stages: first ask pupils to watch the proof. Then later ask them to write it out for homework. Still later ask them to watch it done again. After that they should try writing it out on their own. With some pupils, teachers may find that, given thus in repeated lighthearted doses, the story will both make sense and be remembered with pride.

In the method given in the *Pupils' Text*, which uses similar triangles, draw *two separate sketches*: a picture of the orbit, with vectors to show the velocity at two specimen points, and then a separate vector diagram on which the change of velocity can be shown.

It is helpful to draw the vectors much longer than the radius of the orbit; this will avoid confusion.

Physicists often impose the vector diagram on the main picture. They produce the second tangent vector backward to cut the first vector and then make that intersection the apex of a vector triangle. But with beginners this could make the story much more confusing. In teaching, we need to emphasize the distinction between lengths such as R and AB, and velocities such as v.

A second method, which uses the property of segments of crossed chords of a circle, is given in the form of a structured question (**Question 30**). If pupils know the theorem, this is probably a better method. It follows almost directly from the orbit-drawing experiment. It is, in fact, Newton's own method.

The similar-triangles method As in *Pupils' Text*:

★　　★　　★　　★　　★

An object (or a point) moving round a circle with constant speed v has an acceleration v^2/R towards the centre. Since this is an important expression for many motions in physics, you should see how it is arrived at, and not just accept it as a mysterious 'formula'.

Draw a circular orbit, with an object moving from A to B in time t. The object's speed, along the curved path, is v. At any instant, its velocity is v *along the tangent*. Draw a vector AP to represent the object's velocity at A. That is along the tangent at A. Draw another vector, BQ, of the same length, to represent the object's velocity at B along the tangent there.

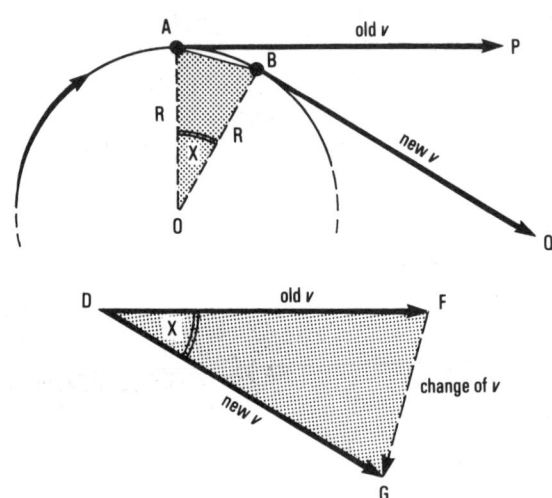

Draw those vectors again in another place nearby; this time *start both from the same point D*. There you have two vectors, each of length v, which we label:

'OLD VELOCITY'　and　'NEW VELOCITY'
(velocity at A)　　　(velocity at B)

Since you want to find an acceleration, you need to know the *change* of velocity. Ask yourself: '*What must be added, as a vector, to the OLD VELOCITY to get the NEW VELOCITY?*' It is the vector FG in the sketch.

Join A and B, and draw radii OA, OB. Then you have two similar triangles, AOB and FDG. That is because each velocity vector is along a tangent, so it is perpendicular to the corresponding radius – so the angles at O and D are equal. Then, from the property of similar triangles:

$$\frac{\text{CHANGE OF VELOCITY}}{\text{VELOCITY, } v} = \frac{\text{chord AB}}{\text{radius } R}$$

$$\therefore \text{ CHANGE OF VELOCITY} = \frac{(\text{AB}) \times v}{R}$$

Suppose this kind of change of velocity, which is perpendicular to the actual motion, is related to an acceleration just like any other acceleration – a surprising supposition, which must be tested. If the supposition is safe, you can calculate that acceleration as usual:

$$\text{ACCELERATION} = \frac{\text{CHANGE OF VELOCITY}}{\text{TIME taken, A to B}}$$

$$= \frac{(\text{AB}) \times v/R}{\text{TIME A to B}}$$

$$\text{ACCELERATION} = \frac{v}{R} \times \frac{\text{(AB)}}{\text{TIME taken}}$$

$$= \frac{v}{R} \times v, \text{ because } \frac{\text{(AB)}}{\text{TIME}} \text{ is SPEED } v.$$

$\therefore$ ACCELERATION $= \dfrac{v^2}{R}$ and is directed towards the centre.

★　　★　　★　　★　　★

{**Proceeding to the limit** In such a discussion, we have the problem of proceeding to the limit as B approaches A. If we do not proceed to the limit, we are left with an approximation, essentially that of calling the *arc AB* approximately equal to the *chord AB*. We really want the acceleration 'at an instant' when the object is at a point A and B combined; so we really want the limit.}

{To keep our calculation in good form, we should put in, at an appropriate place, a factor (*arc AB*)/(*chord AB*); then we should show carefully that this factor tends to 1 in the limit.}

{However, taking that much care would be the last straw for our young pupils, except for those who are natural mathematicians. So we should avoid labouring, or even referring to, the need to proceed to a limit, or the method of doing so. If a pupil objects, we should just point out that the jump we have made becomes more and more trivial as we move B closer to A.}

{**Calculus** There are some quick methods that use calculus. At this stage, such methods are likely to be too obscure, however quick, and should be avoided. We should also avoid methods that use trigonometry: either the sine that is used cancels out – and similar triangles would have been clearer – or the method involves concealed differentiation.}

{The similar-triangle method is closely related to the hodograph method, but we do not advocate the latter because it seems more sophisticated to pupils.}

Putting a $= v^2/R$ to use As soon as pupils have arrived at $a = v^2/R$ we should put it to use. The simplest use that looks real – and not artificial like problems that ask for tension in the string of a whirling stone experiment – is the calculation of the orbital time of an Earth satellite a short distance up. As in the graphical method, point out

that the acceleration of the satellite is much the same as that for a projectile slightly nearer the Earth's surface. So the acceleration is g. Write $v^2/R = g$, take R just over 6400 km (say 6600), and ask pupils to calculate the time of going once round the Earth.

EXPERIMENTAL TEST OF $F = mv^2/R$ FOR CENTRIPETAL FORCE

Pupils should give the new expression an experimental test. Even if they find the geometry and algebra easy, pupils do not feel quite happy about applying $F = ma$ to this motion where the acceleration is *'across the motion' never changing the speed but only changing the direction of motion*. So they should make a test of the prediction that motion in a circle needs a force mv^2/R.

The test: whirling an object pulled by gravity load The pupil whirls a small object round his head in a horizontal circle by a string which passes through a glass tube held in his hand and carries a load hung on its end below the tube.

The load might be replaced by a spring balance anchored to the floor and read by another pupil squatting beside it. That would avoid bringing gravity into the discussion at this point.

Although this simple experiment needs two pupils to co-operate in making measurements, make sure that each pupil takes his turn in doing the actual swinging. To a young person, participation makes the test much more real.

The team measures the time for a counted number of revolutions. The radius of the orbit is measured from the satellite to a mark on the string which is kept just at the mouth of the glass tube during swinging. The force actually used is the WEIGHT of the load hung on the lower end of the string. The pupil compares that force with the THEORETICAL FORCE mv^2/R.

Description of equipment for pupils' test of mv^2/R

This offers a rough test of a 'theoretical formula'. A simple assembly of ordinary materials will be very good for pupils. Not only is a polished commercial version unnecessary but it might give pupils an idea that the formula only applies when special apparatus is used. (Cf. the danger of dry ice pucks leading to the idea that Newton's Laws only apply when there is no friction.) Hence the following notes for home-made equipment.

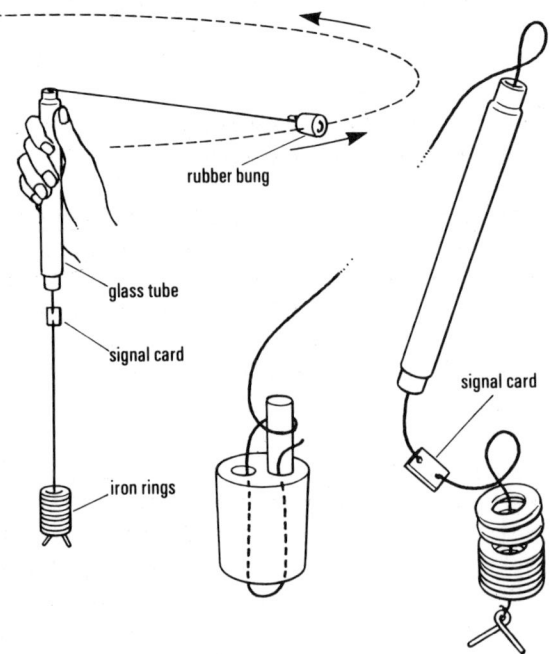

SATELLITE This is a rubber bung 2 to 3 cm in diameter, $2\frac{1}{2}$ cm long, with two holes. A short bit of wood dowel fits into one hole to anchor the cord there. (A block of wood could be used, but a large enough mass would be clumsy.)

LOAD This pulls with a measurable force. It may be any chunk of metal, with a mass of 100 to 200 grams; but a pile of large iron washers is convenient, and it allows the pupils to change the force.

TUBE This, with a pupil's hand holding it, is the driving motor that maintains the orbital motion of the satellite. It is a glass tube, 15 to 20 cm long, 3 or 4 mm bore, with the ends fire-polished so that the cord that runs through it can slide easily. To ensure an easy grip, the tube may be covered with rubber tubing; but if that is difficult to obtain or install a binding of thick plastic tape will suffice.

CORD This runs from the pulling load up through the glass tube and out almost horizontally to the satellite. This may be thin string but a nylon line, plaited or monofilament, slides even better on the glass tube's mouth. Buy at fishing tackle shops.

WIRE HOOK Anything that will prevent the loads falling off the lower end of the cord will suffice. A single loop of stiff wire does well.

SIGNAL CARD To keep the satellite at a constant (measured) radius, the experimenter watches a small signal card on the cord, and tries to keep that just below the tube. A 2-cm square of card, with two holes to take the cord, does well. A paper-clip is less convenient.

Class Experiment 6
Experimental test of $F = mv^2/R$

Aim To give pupils personal experience of centripetal force being estimated by a formula.

Apparatus

From centripetal force kit†	item 172
20 glass tubes, with rubber tubing	172A
20 rubber bungs (satellites)	172B
20 wire hooks	172C
200 metal washers (for loads)	172D
20 Signal cards (e.g. 2 cm × 2 cm)	172E
1 ball of thin cord, preferably nylon	172F or 10A
16 or more stop watches or stop clocks ‡	507

† *For home manufacture*, see the notes on pp. 25–6.
‡ Some pupils might use a wall clock with a seconds hand.

Pupils work in pairs.

Teachers may wish to save time by assembling the devices beforehand, since *in this experiment* the main point is not constructing the apparatus but making an important test with it. However, the preparation described below is repeated in the instructions in *Pupils' Text*.

Preparation

Tie a piece of cord about 1·5 metres long to the wire hook. Pass the other end through the two holes of the signal card; then through the glass tube; and then through one hole of the rubber bung. Then pass the other end back through the other hole of the bung and anchor it by plugging the hole with the wooden rod. Finally hitch the string round the wooden rod, or just tie it there.

Procedure

Pupils follow these instructions.

The experiment Slip several iron rings over the hook as loads to provide the inward pulling force.

Adjust the signal card so that it will be just below the glass tube when the bung is out at the radius you choose for the orbit.

Whirl the bung round above your head, holding the glass tube in your hand. Keep the signal card just below the glass tube. The card will probably spin when it is not touching the tube and this will tell you if it is clear of the tube.

Practice keeping the orbital motion going evenly. Then continue while your partner times 50 revolutions with a stop clock. At the start give him a countdown: 5–4–3–2–1–; then start at 0–1–2–... and go on till 48–49– and stop at 50.

Then change with your partner and let him whirl the 'satellite' while you time 50 revolutions. Take the average time for a revolution.

Calculate the forces Measure the radius R of the orbit, in metres. The signal card will help you to do that.

You will need the mass of the bung, m, in kilograms. Find that using a balance.

Calculate the speed of the bung round the orbit, v. Then calculate the THEORETICAL (predicted) INWARD FORCE mv^2/R.

Work out the ACTUAL FORCE due to the load W. (Remember that the Earth pulls 10 newtons on each kilogram.)

The test You wish to test whether $F = mv^2/R$ successfully predicts the force needed to hold the bung in its orbit. This THEORETICAL FORCE came from thinking about the motion with the help of Newton's Second Law. The *actual force* that keeps the bung in orbit is the inward pull of the cord.

Are these equal? That is the test question.

Although you may hope for close agreement, you will know, when you have done the experiment, that this is rather a rough check. So it is also a test of your own experimental skills.

COMMENTS ON TESTS OF $F = mv^2/R$

What is being tested? Again and again in every discussion with pupils, we need to emphasize the order of reasoning in the test, to avoid letting the logic be reversed. One might say to pupils: 'The *actual* force is the pull of the string, the pull that you measure. The force that you calculate by working out the value of mv^2/R is the *theoretical* force, the force that you hope will be a good prediction of the actual force. You are doing the experiment to find whether your hope is reasonable. It is mv^2/R that is under question, not the actual pull which you know is true.'

To young pupils who have just been carried through a derivation, mv^2/R seems so important that they are apt to insist that it is the real force and the directly measured value is only a blundering attempt to approximate to what they know is really true! Unless we can straighten out that confusion gently but firmly, the test may do more harm than good.

Apparatus for test: a plea for simplicity Ingenious physicists have devised many forms of apparatus for carrying out such a test. It is a tempting problem for all of us; there is a strong need for a test and there are intriguing opportunities for ingenuity and skill; so we devise apparatus for the test and then make it more complex by adding improvements. But when pupils try such apparatus the intrinsic difficulty of the essential idea at stake makes the complexity most unwelcome.

So we should try to keep the apparatus as simple as possible, provided it can yield some kind of quantitative test. We need a simple device in which a known mass is held in an orbit of measured radius with measured speed by an inward force which pupils can measure directly and compare with the calculated force $F = mv^2/R$.

A simple form of apparatus is suggested here. Where a teacher has already devised and made good apparatus of his own for this test, he should certainly use it. The enthusiasm and confident knowledge of the man who made the apparatus are very valuable: they help to carry the pupil through the experiment with enjoyment.

However, where the teacher decides to use our suggestion we urge him to resist the temptation to add improvements, because this experiment is so easily obscured by its own machinery.

Whatever device is used, we hope that it will be simple and cheap. An expensive device would have only limited use as a class experiment and a complicated one will obscure the real nature of the test. Even if the simple cheap one is too rough in its behaviour to afford an accurate test, pupils are now able to distinguish an experimental result that is 'wrong' – that is, contrary to expectation – and one that is right but clumsy.

In carrying out tests in a class experiment we must be careful not to let the test grow too heavy so that it seems to pupils more important than the uses of mv^2/R to deal with electrons, satellites, and the whole solar system.

A systematic investigation in which the force is measured for a series of different orbit-radii, all at the same speed, then for different speeds, and so on, may seem tempting to well-trained physicists, but it would lead our young beginners into discouragement rather than keen understanding.

It may even be better to go ahead to uses of mv^2/R and fit the test in later when there is a good opportunity. Pupils should now be at an age when they understand that the intermediate stages between the general idea of an object moving round a circular orbit and the final result (that the acceleration is v^2/R inwards) are not mysterious pieces of abstruse science or mathematics but consist of ordinary geometry and algebra. Even if the whole story of that connection is kept in a black box, they should know that the box contains only the gears and levers of algebra and geometry that they have met before. If they have gone through some earlier derivations (like Galileo's geometrical derivation of $s = ut + \frac{1}{2}at^2$) these should serve as assurance to enable them to take the new derivation for granted. We should not have to assure pupils that we are honest; but we do have to assure them that they are not misled.

Questions Pupils are now ready to make some calculations such as: the strength of a string to whirl a stone; the period of an Earth satellite near the Earth; the minimum speed for an aircraft to loop the loop; the equivalent 'g' in a centrifuge. Two reminders:

1. When you calculate mv^2/R, with m measured in *kilograms*, v and R in units involving *metres* and *seconds*, remember that the result will be measured in *newtons*, since it is a FORCE.

2. You will often be given the values of radius R and period (orbit-time) T for a circular orbit, but not the value of v which you need for v^2/R. Remember: speed v is circumference/time, $2\pi R/T$.

Calculation for a communication satellite

The general idea of a communication satellite is now well within the scope of our programme. (What does it do? Where is it put? What must be its period?) But the calculation of its necessary period – 24 hours – is outside the normal scope of our programme, solely because of its cumbersome arithmetic.

FURTHER USES OF mv^2/R

Pupils will use their ability to calculate the inward force needed to maintain an orbit, for electrons etc., and for the Moon's motion, and for our Newtonian study of the solar system. In each case, they need some new knowledge before we can put the central force expression to use. They need: the force on charges moving across a magnetic field; the inverse-square law of gravitation; and some historical knowledge of planets, etc.

Since those uses cannot be shown at once without such extra knowledge, we choose electrons first because we can continue the work of Year 4 with them.

CHAPTER 2

Measuring electrons

Programme In the work of this chapter, pupils use the knowledge they now have to measure e/m for electrons – and their speed from a 100-V gun. They then compare the result with e/M for hydrogen ions. Pupils see the electron as a chip off an atom, with most of the atom's mass in the remainder – a positive ion. A short description of mass spectrometers follows, with a discussion of their use.

At this stage we offer neither experimental support nor compelling need for a nuclear atom model; so, to maintain scientific honesty, within this chapter we keep the simple 'pudding' model of an atom – electrons embedded in a positive core, like currants in a bun.

Preliminaries Since the main experiment of this chapter uses the 'catapult' force of a magnetic field on a stream of electrons, we suggest that the teaching should start with a further demonstration of the fine-beam tube and a revision of 'catapult' force effects.

DISCUSSION OF THE ELECTRON GUN

The Teltron fine-beam tube, with the horizontal gun used to fire a stream of electrons across to the screen, offers an opportunity to consider the action of the electron gun. Explain that electrons leave the heated cathode with very small speeds, but we apply an electric field which accelerates them towards the plate. (We must make it clear that the p.d. we apply to accelerate the electrons – the gun voltage – is nothing to do with the supply used to heat the filament or cathode.)

Unlike electrons in a wire, these electrons have nothing to hit, nothing to give energy to, so they travel faster and faster across the vacuum to the plate. All the energy they gain must be retained as K.E. until they reach the plate. Those electrons that hit solid metal are brought to a stop, and share out their K.E. as heat among the atoms of the metal. (Any X-rays? At most, only a very tiny fraction of the collisions produce X-rays, even in the best of X-ray tubes!)

Suppose each electron has a charge e (measured in *coulombs*) and a voltmeter connected between cathode and plate shows a p.d. V (measured in *volts*, which are *joules per coulomb*). Then the supply gives energy V (joules) to each coulomb passing from cathode to plate. Therefore each electron gains energy Ve (joules); and that is K.E.

Those electrons that arrive at the 'gun muzzle', the hole drilled in the plate, keep their K.E. and go on through the hole. After that (unless we add a further battery) there is no more acceleration: they keep a constant speed until they hit some barrier.

Pupils will need to have a clear idea of such an electron gun, from which electrons emerge in a stream from the muzzle, and thereafter continue with unchanging speed. And they need to understand why we say the K.E. of each such electron is equal to its charge times the gun voltage: $\frac{1}{2}mv^2 = eV$.

Notes on the fine-beam tube

The fine-beam tube has been used in earlier Years, but in this Year it takes on new importance because it is used in a measurement of e/m for electrons. Therefore we recapitulate here details of its adjustments and behaviour.

We suggest using the Teltron tube, in which the electron stream is marked by a green glow in low-pressure helium. The tube has the advantage of two guns – one firing along the axis, one perpendicular to it. Schools that already have a Leybold tube should of course use it, following the instructions they will have for it.

The tube has two guns. There is a two-way switch at the rear of the tube which connects the supply to one gun or the other. In each gun the cathode is heated independently by a small heater.

Just outside the muzzle of each gun there are two slanting plates which can be used to provide an electric field to deflect the beam. The 'outermost' plate of each pair is attached to the gun muzzle (anode). The other plates of the two pairs form a single V-shaped piece of metal which is connected to a plug on the outside of the tube.

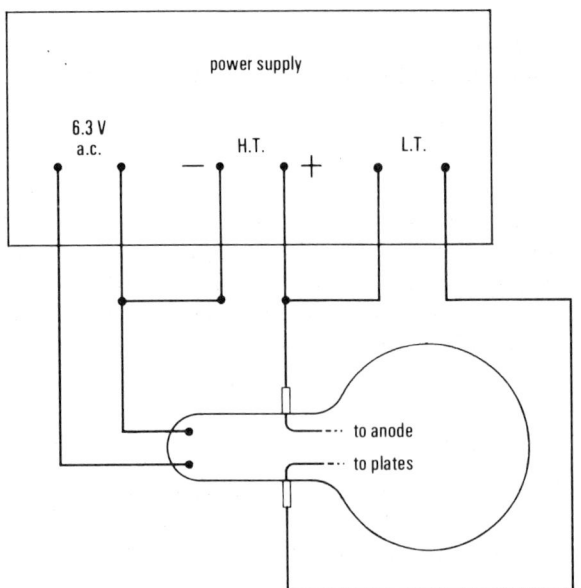

The following are labels within the diagram:

power supply

6.3 V
a.c.   H.T.   L.T.
         −   +

to anode

to plates

The beam may focus more sharply if a small p.d. is applied to the deflecting plates.

Setting up and running the tube The tube is designed to be held in the same stand as other Teltron hot-cathode tubes. Before installing connections, place the tube in its stand.

When Helmholtz coils are to be used to make a uniform magnetic field across the tube, set up the coils on the stand *after* the tube has been placed in position. Connect the coils in series, with a battery (6 to 12 V), a 15 Ω rheostat and ammeter (0–1 A). (An ordinary power supply may not do so well, if it provides only full-wave rectified a.c., whose bumps will make the deflected beam less sharp.)

Start up the tube by running the heater without any gun voltage. With all power supply potentiometers set to zero raise the heater voltage to provide about 0·3 A.

Then raise the gun (anode) voltage. At about 80 V, a short green beam will appear. As the gun voltage is increased further, this beam will lengthen and will strike the wall of the tube. (As with most gas tubes, once the beam is established it will continue even if the gun voltage is reduced below the striking value.) Use 80 to 120 V for the gun.

When a uniform magnetic field is needed, switch on the current in the coils. The beam will be bent into part of a circle. Raise the current. A current of about 0·2 A in the coils will hold the beam in a complete circle. If the orbit looks like a

spiral, twist the tube slightly in its holder until the orbit looks like a circle.

BENDING THE ELECTRON STREAM IN A FINE-BEAM TUBE

Demonstration 7
Electron stream in a fine-beam tube;
effect of a magnetic field

This should be a light-hearted preliminary demonstration, yet an important one. Although pupils have seen the fine-beam tube before, they now have serious use of it ahead, and they should see the effect of a magnetic field on the beam at close quarters – a remote glimpse would make a very poor beginning.

If pupils stand close to the tube they can see the beam well even if the region round the experiment is only half dark; the rest of the room can be well enough lit for other pupils to read or write – as long as there is no direct sunlight.

Pupils should come to the experiment in groups of four or six and watch how the beam behaves when it is tethered by the field of a single magnet – a single coil carrying current.

This is a qualitative demonstration, measurements will come in a later experiment.

Apparatus

1 Teltron fine-beam tube	item 235(61)
1 stand for Teltron tube	140
1 of a pair of Helmholtz coils for the tube	236(62)
1 H.T. power supply with 6·3 V supply for the heater	15
2 rheostats (10–15 Ω) (one for the heater, one for the coil)	541/1
1 12 V battery	176
2 Magnadur slab magnets	92B
1 bar magnet	
1 5 kΩ resistor, 10 W (safeguard for tube anode circuit)	
1 voltmeter, 200 or 300 V d.c.†	
2 ammeters (1 a.c., 1 d.c.), 1 A	

† The meters are used only for making adjustments and need not be demonstration instruments for pupils to see. The voltmeter may be incorporated in the H.T. power supply; otherwise use item 70 with 71/11.

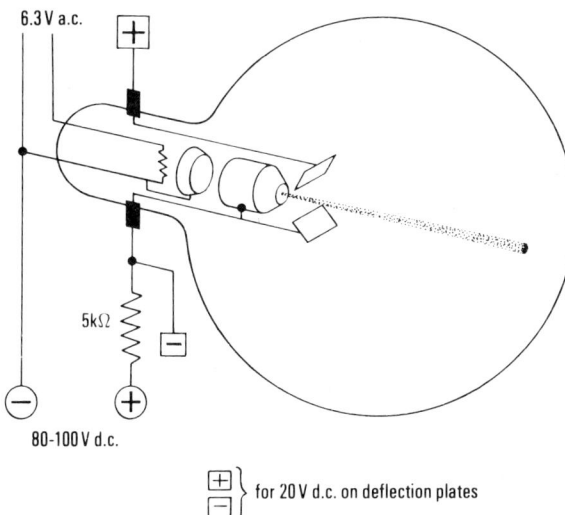

6.3 V a.c.

5 kΩ

80-100 V d.c.

for 20 V d.c. on deflection plates

Procedure

See the notes on the Teltron tube in the box on p. 27.

Set up the tube and connect *both* deflecting plates to the anode.

Switch on the heater of the gun which fires a horizontal beam. Adjust the current to 0·3 A. Wait until the cathode is heated.

Raise the gun voltage (anode voltage) until the beam hits the end of the tube. Use 80 to 120 V for the demonstration.

(*i*) Bring a bar magnet near the tube while the pupils watch.

(*ii*) Bring a magnet with face polarity (a Magnadur slab magnet – item 92B) near the tube taking care to avoid sharp contact with the glass.

(*iii*) Without installing the coils in the tube-base, turn on a current of about 0·2 A in one of them and carry this near to the tube.

The demonstration shows that the beam is bent most where the magnetic field is strongest, and that the deflection is at right angles to the motion of the electron stream.

Deflection by electric fields An electric field applied across the stream of electrons will deflect the stream. Pupils will see the beam bent towards the *positive* deflecting plate. They will infer that the particles in the beam carry *negative* charges, *whichever way they are moving*. They will have to find which plate is positive by tracing connections from the battery or power pack.

Demonstration 8
Electric field deflects an electron stream

Apparatus

As for Demonstration 7, without the equipment for magnetic fields. The voltage for the electric field may be provided by two 12 V batteries or by the 0–25 V d.c. supply of the power pack if it is well smoothed.

Procedure

Use the tube as set up for **Demonstration 7**.

Throw the two-way switch at the end of the tube so that the gun firing a horizontal beam is used.

Explain that each gun has one of its deflecting plates attached to its own anode (muzzle). The other deflecting plates, one for each gun, are joined

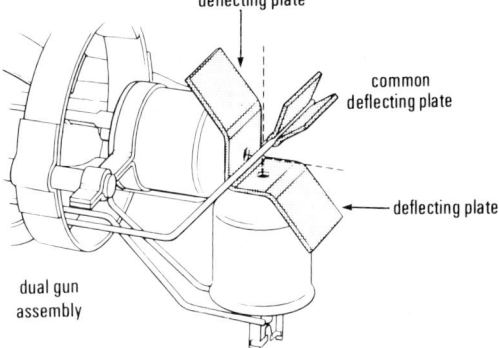

deflecting plate

common deflecting plate

deflecting plate

dual gun assembly

together to form a V-shaped piece of metal sheet, which is connected to the external socket for a deflecting voltage.

Apply a p.d. of about 20 V to the deflecting plates. Connect the +ve of the battery to the external socket, and the −ve to the anode connection. Then ask: 'Are the things in the stream charged + or − ?'

CATAPULT FORCE

In Year 3 pupils tried experiments to see a magnetic field exerting a force on a wire carrying current across it. The force could be illustrated by combining the magnetic fields of the current and the magnet. The lines of the resultant field could be imagined to be stretched elastic strands, trying to catapult the wire in the actual direction of the force. We gave qualitative descriptions of the use of such catapult forces in motors, ammeters, and loudspeakers.

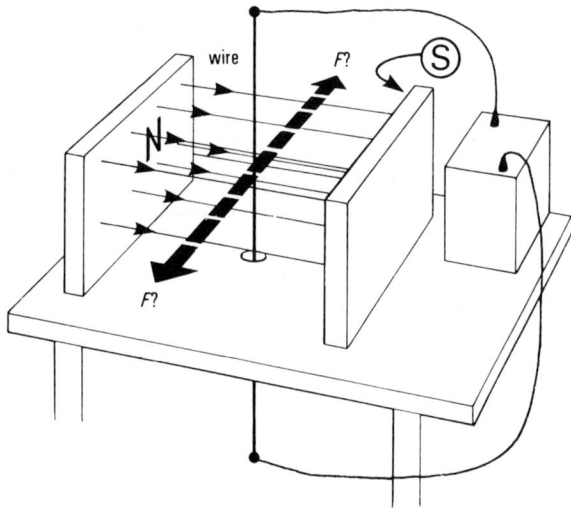

Force *F* on the wire is in or out, perpendicular to the paper.

Now in Year 5 pupils will see streams of electrons held in circular orbits by catapult forces; and they will make an important measurement. We must give pupils a quantitative account of catapult forces, and we must transfer the expression from *force on a current* to *force on a moving charge*.

To make sure that pupils remember their acquaintance from Year 3 and to provide for newcomers who missed it, we repeat some Year 3 sketches here and in *Pupils' Text*.

They will need a vivid feeling for these forces. As a reminder, take them back to the electromagnetic kit and ask them to repeat the experiment with a movable wire carrying current across a magnetic field. However clearly they seem to

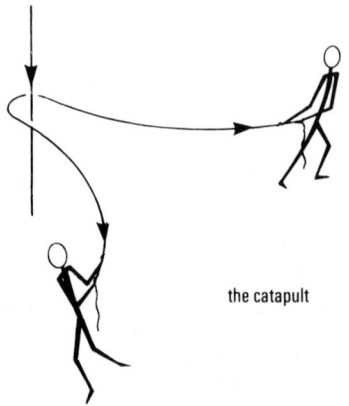

the catapult

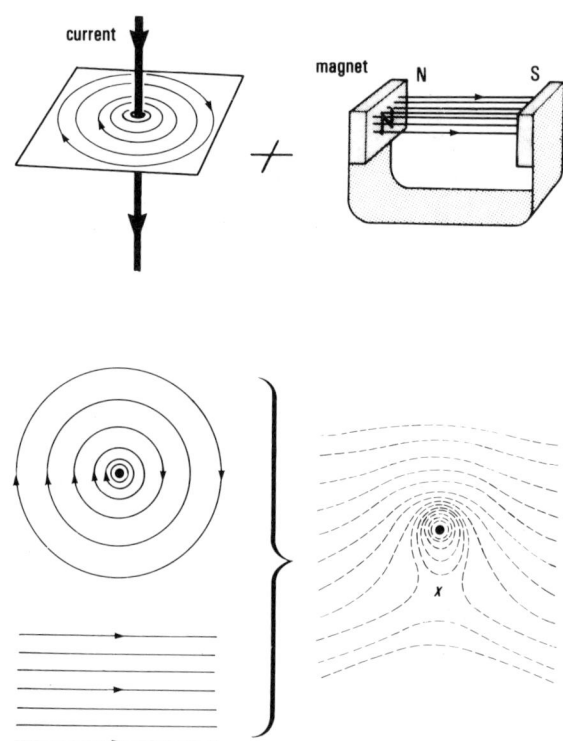

remember it from the earlier Year, they should try it again.

Any who have not seen the iron filings map of the catapult field deserve to see it now. For that, repeat **Demonstration 96a** of Year 3. Some pupils might now set up **Experiment 95b** as a pupil demonstration for revision.

Class Experiment 9
Movable bridge and the catapult force

Apparatus

From electromagnetic kit 16 sets of:	item 92
iron yoke	92J
pair of Magnadur slab magnets	92B
PVC-covered copper wire	92X
copper wire, bare, SWG 26	
16 wire strippers	84
16 *pairs* of support blocks★	219(=92CC)
16 low voltage d.c. supplies	104

★ These may be needed if the terminals of the d.c. supplies are not suitable as supports.

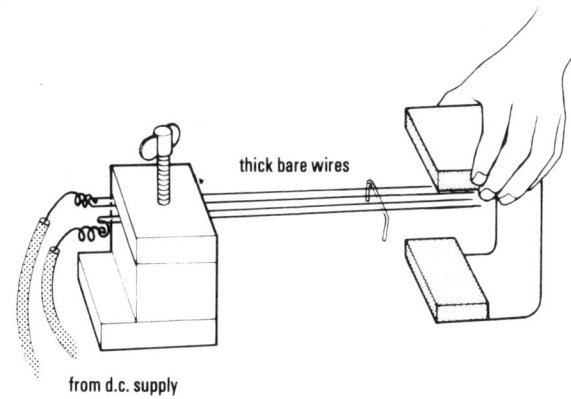

thick bare wires

from d.c. supply

Note It is helpful if the ends of the two rails are turned up to act as buffers and if the ends of the wire bridge are turned down as shown.

Procedure

Pupils work in pairs and follow these instructions:

    ★   ★   ★   ★   ★

Clamp two 15-cm lengths of thick, *clean*, bare copper wire in a support block as shown. Use a third piece of wire to make a movable bridge resting on these two rails.

You may know from an earlier experiment how to place the U-magnet. Place it so that you expect it to make the bridge move when you switch the current on or off. If you are not successful try other positions for it. What happens when you reverse the current?

Electrons in orbit Since the catapult force is perpendicular to a stream of charged particles, a *uniform* magnetic field will hold the stream in a circular orbit. We shall make use of this property in our measurement of e/m of electrons. As a preliminary, pupils should now see the electron stream in a fine-beam tube pulled into a circular path.

For a bonus of delight, twist the tube slightly in its holder. Then the circular motion combines with a linear component to make a spiral.

We may explain, as in the *Pupils' Text*:

    ★   ★   ★   ★   ★

Now make a *uniform* magnetic field – with the same strength and direction across the whole tube. That can be done by sending a steady current through a pair of coils* suitably placed. Then the catapult field has the same steady value at all places along the stream, as long as the electrons keep the same speed.

Since the catapult force is always perpendicular to the stream (as it is when it acts on any electric current), it does not change the speed of the electrons. It only pulls their path into an orbit. It acts as a tether, like a string that holds a whirling block in orbit. With that constant catapult force, the orbit is a circle.

Demonstration 10
Fine-beam tube; effect of a uniform magnetic field

Apparatus

As for Demonstration 7, without the bar and slab magnets.

Procedure

Set up the tube and the Helmholtz coils on the stand. Connect the coils in series (pointing out the

*A current through a Helmholtz pair of coils makes a remarkable uniform magnetic field in the region between the coils *near the axis*. The arrangement was originally designed for use with a short magnet near the axis. But the orbit of the electron stream is a considerable distance out from the axis –

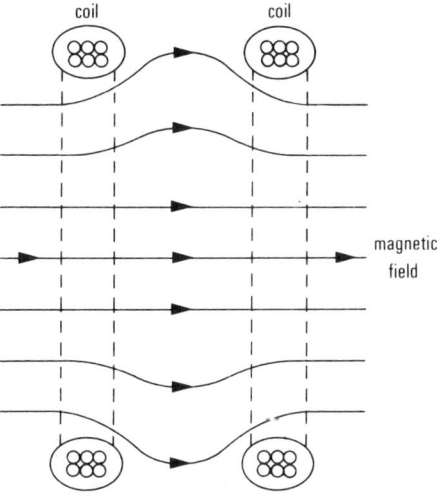

coil   coil

magnetic field

These are the traditional Helmholtz coils, spaced one radius apart along their common axis.

say, $\frac{3}{4}$ of the coil radius for small coils – and there the field is not given quite accurately by the usual axial formula. Fortunately, we do not use the formula but estimate the effect of the field directly by substitution. Otherwise we might have one more small uncertainty.

31

connection linking the two) with a battery, a 15 Ω rheostat and an ammeter (0–1 A).

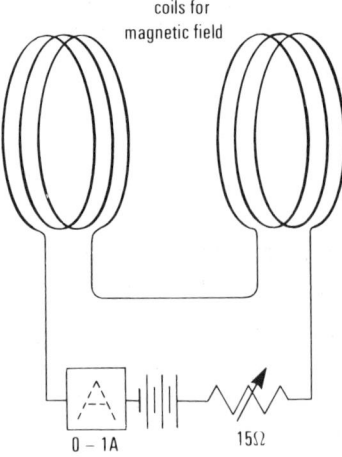

coils for
magnetic field

0 – 1A   15Ω

Use the switch at the end of the tube to select the gun that fires a vertical beam.

With all the power supply potentiometers set to zero, raise the heater voltage to provide about 0·3 A. Wait till the cathode is well heated, then turn on the anode (gun) voltage and raise it until the beam strikes the wall of the tube (80 V or more).

Let the pupils see the straight, vertical beam.

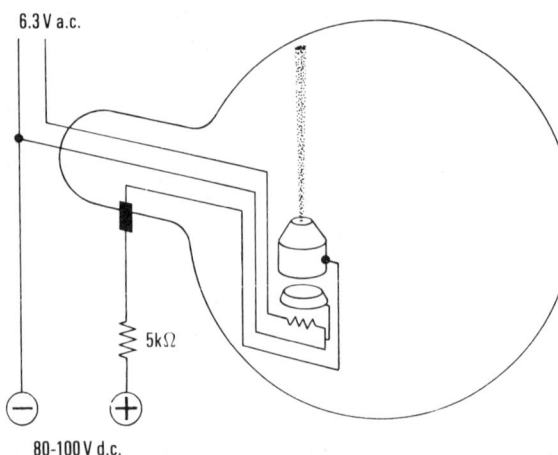

6.3 V a.c.

5kΩ

80-100 V d.c.

Teltron fine-beam tube with vertical gun in action

Switch on the current in the coils for the magnetic field. Increase that current until the stream forms a complete circle. A current of 0·2 or 0·3 A will do this. Adjust the current and the gun voltage (80 to 120 V) for best visibility.

If the beam has a spiral path, twist the tube slightly in its holder.

The beam may focus better if a small p.d. (1 or 2 V) is applied to the deflecting plates.

If time permits show what happens when the gun voltage is changed and when the current in the coils is changed. But the most important use of the time is to let the pupils come in small groups close to the tube, to see clearly the straight path and then the orbit.

MEASUREMENT OF ELECTRONS:

v and e/m

Pupils should use the fine-beam tube, taking turns in very small groups. This is a very important 'atomic' measurement and it should be treated partly as a class experiment. The teacher should assemble the apparatus and obtain a beam. Then pupils should make a measurement.

Measurements In our experiment, we make measurements with an electric field and with a magnetic field. (See Notes and Comments, later, for the necessity of both measurements.)

The electric field measurement should be that of the gun voltage, to be used in Equation (I) below.

The magnetic field should bend the path of the stream into a circular orbit whose diameter pupils measure. This is used in Equation (II) below.

Preparing the ground It may be a good thing to run a demonstration right through first. Except with a very fast group, go through the calculation completely in the demonstration but ask the pupils not to make any record of it.

Then leave them to carry out their own calculation when they do the experiment.

This is a grand experiment, perhaps the most impressive one in the whole course in which pupils participate. There is a serious danger of both the doing of the experiment and its result being swamped by the amount of preliminary teaching to be remembered – ideas, definitions, and relationships. To avoid such a disappointing fate, one needs to start with a clear reminder of the aim of the experiment and an offer of considerable revision.

Review the meanings of ELECTRIC CHARGE and POTENTIAL DIFFERENCE; the definitions of *newton, joule, coulomb, volt*; the idea of electric field; the name 'electron gun' and its meaning, the idea of electrons accelerating in an electric field in a vacuum, and of their continuing with constant velocity outside that field.

It is important to avoid the other extreme of over-tedious preparation. This is a matter for judgement in offering revision just when it is needed.

Electric measurement: gun voltage In a hot-cathode tube, such as the fine-beam tube, all the electrons that emerge from the muzzle of the gun have the same kinetic energy, equal to the ELECTRON CHARGE times the gun voltage. Therefore we can write

$$\tfrac{1}{2}mv^2 = Ve \qquad \text{Equation (I)}$$

This measurement of gun voltage provides one of the two pieces of experimental information that we need. The effect of a magnetic field provides the other.

Magnetic measurement A current through a pair of coils one radius apart along the axis makes a fairly uniform magnetic field in the region half way between the coils. Place the coils so that the electrons' orbit is in that field, and let pupils measure the orbit diameter.

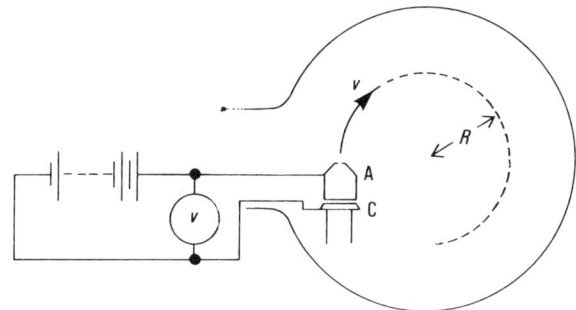

Explain to pupils that the catapult force on a wire of length L carrying current I is given by $F = BIL$, where B is a measure of the magnetic field (or rather flux density). Show them how we argue that (IL) can be replaced by (ev) for a particle with charge e moving with speed v across the magnetic field.

Then, halving the orbit diameter to obtain the radius R, we say:

force *needed* for orbit =
            force *provided* by catapult force

$$mv^2/R = Bev \qquad \text{Equation (II)}$$

We measure the value of B by putting a simple current balance carrying a known current in place of the fine-beam tube, with the same magnetic field maintained there.

From $F = BIL$ to $F = Bev$ The change from (IL) to (ev) is described in *Pupils' Text* but pupils will need some help and reassurance that the details are not 'examination material'. So the discussion is repeated here.

The new current balance is much simpler than the earlier model; it is described here as well as in *Pupils' Text*.

DISCUSSION OF FORCE ON MOVING CHARGED PARTICLES

A formula for the force Pupils already know that there is a catapult force on a wire carrying a current across a magnetic field. Tell them that the force is proportional to the current, and presumably proportional to the length of wire. (A double length of wire, as in the sketch in *Pupils' Text*, p. 19, should vouch for this.) A moving-coil ammeter with visible works shows that force in action; its uniform scale shows that the force is proportional to current, if we trust Hooke's Law for the hair-springs.* We express this knowledge in the form:

force = (B)·(current, I)·(length of wire, L) or
    $F = BIL$

where B is a measure of the magnetic field (or flux density) combined with a constant that is determined by the choice of units.

In our present teaching, we need not discuss the value of that constant or the units for magnetic field, because we use only one magnetic field for measurement. We use it to deflect the stream of electrons in our measurement of e/m; then we estimate the value of B for the *same* field by a direct measurement of the force it exerts on a known current in a known length of wire.

Change from (IL) to (ev)

We need to transfer that knowledge from the case of a current in a wire to the case of a stream of charged particles. This is a very big and difficult step for pupils, and we should comfort them by telling them it was a difficult step for all scientists when the well developed theory of electric circuits and forces had to be extended to moving electrons, etc., late in the last century.

---

* This proportionality of force and current is implied by the definition of current measurement and no experiment could possibly prove it. Even so, the ammeter illustrates it well.

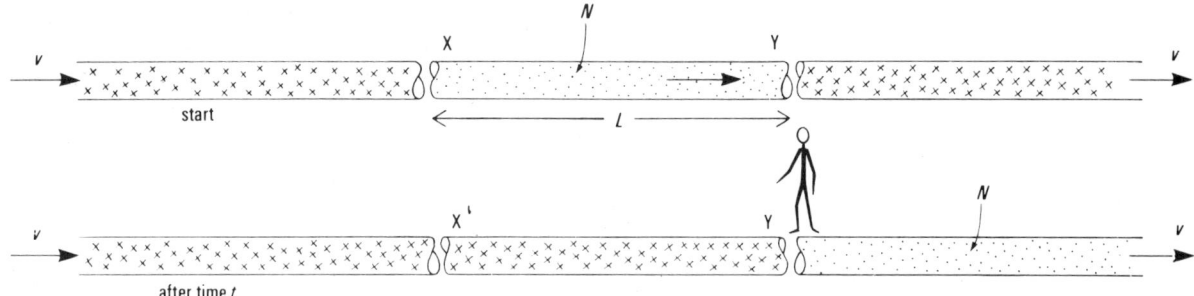

The argument Draw a section of wire XY of length L containing N mobile electrons each of charge e. We suppose that when there is a current I, the electrons move with speed v from X to Y.

Post an imaginary observer at the 'outgoing' end Y and ask him to count electrons like a child counting cars as they go by. He starts counting when an electron emerges at Y. The electron that was at X at that instant arrives at Y some time t later, having travelled distance L with speed v.

In that time t, all the N electrons in the wire between X and Y arrive at Y. Therefore in time t the observer counts a total charge Ne and says the current I is Ne/t.

But vt is the length L.

$\therefore IL = (Ne/t) \times (vt)$ or Nev.

Assume that the force on current I through a length of wire L is BIL where B is a constant, involving the strength of the magnetic field. Then the force is also $BNev$.

Therefore the force on a *single moving charge, just e*, is Bev.

Like the derivation of v^2/R, this sophisticated argument will not succeed with young pupils unless we preface it with two encouragements: (1) show pupils that we and they badly need the result; (2) assure them that this is not a proof that they must learn for reproduction in Nuffield examinations. (Its result will be printed on any Nuffield examination papers in which it may be needed.)

This is like something high up that pupils can only just reach – not inaccessible, but not to be fully grasped. They should enjoy the privilege of touching it, but then they need only remember that what they saw does make sense and is not mysterious nonsense.

CURRENT BALANCE TO MEASURE FORCE DUE TO MAGNETIC FIELD

In all this, the constant B remains unknown. It contains the strength of the magnetic field. Its value also depends, of course, on our choice of units and of 'system of units', which in electricity and magnetism often contains natural constants relating to the properties of materials and even the properties of vacuum.

Here, we shall not go further with the nature or value of B but shall measure its value by a direct experiment on a known sample current in the actual magnetic field that we use for electrons. For that we need a simple current balance.

We suggest a simple design that will weigh the force on a short section of straight wire in the magnetic field of the coils and current used for the fine-beam tube.⋆ See the description in the box.

The new current balance

THE NEW FORM Instead of a rectangular frame rocking on two razor blades, the new form is a simple loop of copper wire which is anchored at one end with the other end free to sag under any applied force.

THE LOOP is an open-ended rectangle of bare copper wire, SWG 26, about 30 cm long and just 10 cm wide. To form the loop, work on a flat surface, holding the wire down with short strips of tape.

INSTALLING THE LOOP Clamp the open ends of wire in two terminals which hold them 10 cm apart on a rod of insulator fixed in a boss on a stand.

⋆ In the first edition our current balance was a loop of wire balanced on two razor blades through which the current was fed in and out. It was not very easy to use. Also there have long been more complicated and robust commercial forms. At this stage we need simplicity and ease of use. So we have now changed to a simple design which we urge teachers to use.

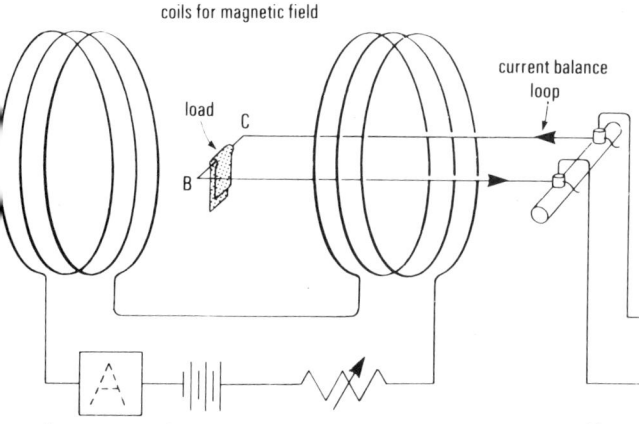
coils for magnetic field

current balance loop

load

C

B

Keep same current

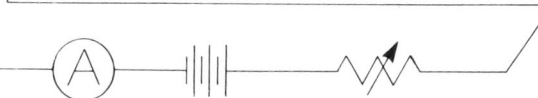
Measure current for same sag

on the end of the loop. Mark the new position to which the loop sags.

The tape used in the timer experiments has a nominal mass around 1×10^{-3} gram per metre. So a 5-cm length will exert a downward force on the loop about 5×10^{-4} newton. *It is essential to check this by weighing a metre of tape before the experiment.*

CATAPULT FORCE ON THE LOOP Remove the tape

Adjust the length of the long sides of the loop to about 25 cm. Connect to the ends beyond the terminals with crocodile clips, or plug into the terminals.

THE BALANCE IN USE Remove the fine-beam tube from the stand. Slide the closed end of the balance loop into position so that the 10-cm end is in the region previously occupied by the electron beam in orbit.

Arrange a marker (a piece of scale, or a pin as a

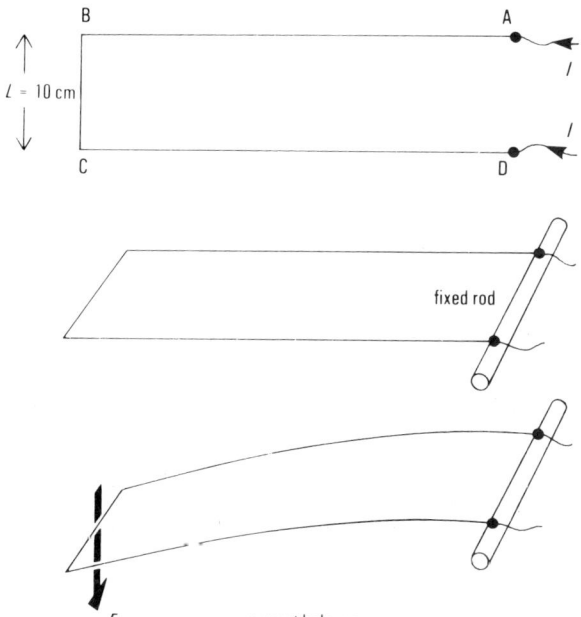
B   A

$l = 10$ cm

C   D

fixed rod

current balance

F

pointer) to register the position of the end of the loop.

Connect the balance in a circuit with a battery, a rheostat (e.g., $1\,\Omega$ with 10 A maximum), and an ammeter 0–10 A.

THE LOAD With both currents off, hang a short piece (say 5 cm) of paper ticker-timer tape as a load

load.* Switch on the current in the coils and make sure it is the same as in the measurement of the electrons' orbit.

Switch on the current in the balance loop and increase it slowly until the balance has exactly the same sag as when paper tape was hanging on the end. Then the interaction between the two magnetic fields is producing the same force on the loop as the weight of 5 cm of tape did. Read the balance current (around 7 A).

Demonstration and Class Experiment 11
Measuring v and e/m for electrons

Apparatus

As for Demonstration 7 but without the bar and slab magnets.
Additional requirements:
1 further 12 V battery†   item 176
1 ammeter, 0–10 A
1 current balance (new form)‡

† Batteries are best for the magnetic field coils and for the current balance. Avoid using a power pack for the field coils unless its output is very well smoothed
‡ Easily assembled afresh when needed.

Preparation

Set up the tube and the Helmholtz coils as in

* Alternatively, keep the tape on the loop and measure the (reverse) current in the loop that brings the loop back up to the original mark. This may be more confusing to pupils; but it may avoid a small error due to elastic changes of the wire, as it is a null method.

Demonstration 10. Connect both the deflecting plates to the anode and select the switch which operates the vertical gun.

Make sure the heater and gun voltage are adjusted to produce a good, clear beam.

Weigh 1 m of paper tape and record the mass for the pupils to see.

Procedure

Follow the general instructions given in earlier demonstrations.

a. Measuring the electron stream's orbit Start up the beam and adjust it so that pupils can see it easily. Turn on the magnetic field current and adjust to give a large clear orbit.

Record the gun voltage, and check it from time to time. Also record the current through the field coils since that must be the same in the current balance experiment.

Bring pupils close to the apparatus, in groups of four to six, to make their own measurement of the diameter of the orbit to the nearest mm. See notes below for methods of making that estimate.

b. Measuring the constant B, using the current balance See the notes on p. 34.

Help pupils to calculate the value of B.

If you wish, offer a *fictitious* example as follows. Suppose the weight of the tape load is 3×10^{-4} newton and the end of the loop is only 7 cm long instead of the real 10 cm. Suppose the current in the loop to match the tape load is 8 A. Then

$$\text{weight of tape load} = BIL$$
$$3 \times 10^{-4} \text{ newton} = B \times (8 \text{ A}) \times (0.07 \text{ m})$$
$$\therefore B = \frac{3 \times 10^{-4}}{8 \times 0.07} = 5.3 \times 10^{-4}$$

for that current balance AND for that orbit.

Note that these data do not fit a real orbit in a real experiment – that is to avoid their being recorded or used by mistake. The values are all prime to each other, to discourage confusion by errors in cancelling.

c. Speed Then ask pupils to calculate v, using some algebra with equations (I) and (II) which they have in *Pupils' Text*.

Teachers, familiar with the history of electron measurements, know that we must proceed to e/m for an essential comparison with protons and they are tempted to carry the algebra through to e/m and obtain a value of that first. But we urge teachers for

reasons given above to proceed to v first. We argue thus:

(II) $Bev = mv^2/R$

(I) $Ve = \frac{1}{2}mv^2$

$\therefore \quad \dfrac{Bev}{eV} = \dfrac{mv^2/R}{\frac{1}{2}mv^2}$

$\therefore \quad \dfrac{Bv}{V} = \dfrac{2}{R}$

$\therefore \qquad v = 2V/BR$ and we calculate v from that.

Results The estimate of v is affected by errors in the current balance experiment as well as those in the electron stream measurements.

Teachers may find it useful to keep some values of v in mind for use as rough checks. The following are correct to within a few per cent:

Gun voltage (V)	Electron speed (m/s)
70	5×10^6
85	5.5×10^6
100	6×10^6
140	7×10^6
180	8×10^6
230	9×10^6

The estimated speed Pause there and comment on the huge value of v for a small gun voltage of a hundred volts or so. Speaking in a sloppy, qualitative way, we might say that this means that e is enormous compared with m.

{Of course, we cannot, as good scientists, compare the numerical values of two utterly different quantities like that. We mean: by comparison with the charges which we can place on large masses, the electron's charge is enormously bigger than we might expect for something of its mass.}

{More striking still, when we know the constants in Coulomb's Law and the Law of Gravitation, we find that the electrical repulsion between two electrons is so enormously greater than the gravitational attraction between them ($10^{40}:1$) that the latter would be negligible.}

d. The very important value: e/m Ask pupils to substitute the value they have calculated for v in one of the equations (I) and (II) and calculate the value of e/m. Substituting in (I), $Ve = \frac{1}{2}mv^2$, is probably the easier.

(Teachers who want to make a quick check of a pupil's result might try estimating $2V/B^2R^2$ – from (II)^2/(I) – which gives e/m without involving a miscalculated result for v.)

Suggestions for measuring the orbit diameter d

Method 1 A pupil holds a ruler outside the tube. In the half-dark room the ruler should be illuminated. A Perspex ruler to one end of which a small electric lamp is taped is suitable. The lamp should be covered with masking tape so that no direct light emerges.

First, let pupils practise on a circle of wire, measuring its diameter from a distance without bringing the ruler near it.

Method 2 Form a virtual image of an illuminated scale inside the tube, in the plane of the electron stream. Place a vertical sheet of clean window glass just in front of the tube. Place an illuminated scale in front of that sheet at such a distance that the virtual image of the scale, formed by reflection at the glass, is in the middle of the tube. This makes measurements easier, but some pupils find it puzzling.

Method 3 Make use of the fluorescent-screen cap of the Teltron tube. Adjust the orbit so that it just touches the plane through the rim of the cap. Then there is a definite diameter to measure, from the gun muzzle to the plane containing the rim of the cap. This very good method has a possible disadvantage: it might reduce pupils' sense of participation. But it has one great advantage: the measurement can be carried out subsequently in full daylight.

Results The value for e/m is 1.76×10^{11} coulombs per kilogram. The most difficult measurements to make accurately are the orbit diameter and the current balance force. (Even if we replaced the latter by magnetic field strength calculated from current and coil dimensions, we should still expect an uncertainty of several per cent.) So we should not claim this as a very precise measurement in any case.

Rough but very important measurements Here, with our rough measurements, errors up to 40 per cent may well be expected. A large error *is* disappointing. To avoid spoiling the success of the experiment by a feeling of disappointment, teachers are urged to discuss the matter with the class *beforehand*. One might say, 'This will be a rough measurement. It will not be accurate because we cannot get inside the bulb and measure the circle precisely, and our current balance will make a rather rough measurement.

'But the result will be real knowledge of electrons. You will find out how fast they move from this gun, also the ratio of ELECTRIC CHARGE to MASS for each single electron. The experiment will be rough, but worth doing.

'Suppose the speed for a small voltage on the gun were really 100 kilometres a second, and our rough measurements gave 70 km/s in one experiment, 130 in another, or even 200. You would still have a very useful idea of the speed of electrons and their charge/mass value. In exploring the sub-atomic world, you would all be in the right county. Do you want to try this, knowing it is rough?'

Importance of e/m measurement While the value of e/m is to us an intensely interesting piece of atomic information, it may seem to pupils a dull thing to work out unless we advertise its importance by pointing out two things in retrospect.

1. This is a piece of information about extremely small things, individual electrons, information that we obtained from large-scale measurements. We never applied a microscope to our experiments. We never counted some vast number of electrons or alpha particles or anything like that. We made ordinary-sized measurements and obtained atomic information.

2. We can compare this measurement of e/m with e/M measurements for other 'atomic' particles: the ions that carry currents in solutions. (And we should assure pupils that ions in gases have similar values of e/M.)

Electron *vs.* atom At this point it is very important to put the e/m result to some use. Otherwise pupils may consider that they have been carried through a long, pointless business – somewhat like the custom of deriving $PV = \frac{1}{3}Nmv^2$ without putting it to use, except for an assertion that it 'proves' Boyle's Law. So we should at once show pupils how to compare an electron with the lightest atom, hydrogen.

Then we can go straight on from hydrogen ions to other ions, mass spectrometers, a simple atom model, and (long before we come to a nuclear atom in a later chapter) some mention now of fission and fusion as predictable sources of energy – just on the basis of mass spectrometer measurements.

Hydrogen ions: measurement of e/M Pupils now need to see a demonstration of electrolysis of water with *measurement of the hydrogen liberated* (unless they saw that already in Year 4 or have seen it in Chemistry).

Demonstration 12
Electrolysis of water

Apparatus

1 Worcester gas voltameter kit†	item 54
1 rheostat (10–15 Ω)	541/1
1 d.c. ammeter (0–1 A)	79
1 12 V battery‡	176
clock	
distilled water with about 2% H_2SO_4	

† The 250-cm³ burettes are very long. If a sufficiently tall jar is available, so that a burette can be pushed down into the jar until it is totally immersed, it can be filled simply by opening the tap. But since that requires an unnecessarily tall jar the kit includes a device for filling a burette: a plastic bottle that can be squeezed.
‡ Instead of the 12 V battery and rheostat, the L.T. variable voltage d.c. supply could be used.

Procedure

Connect the circuit as shown.

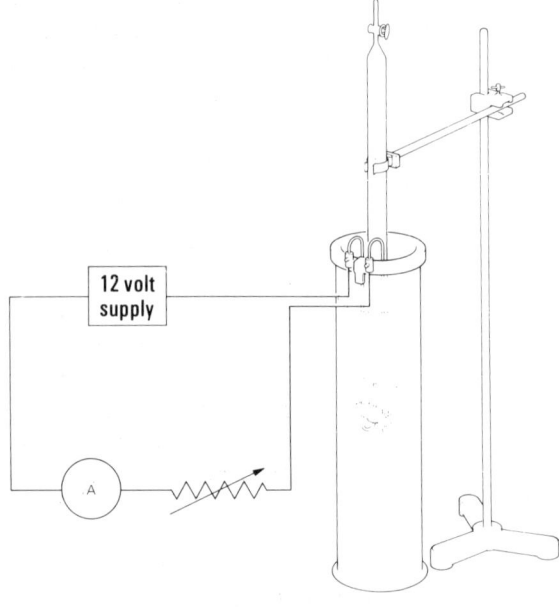

12 volt supply

Only one of the two 250-cm³ burettes is needed. Fill it with acidulated water after it is placed over the cathode. The filling is done as follows:
(*i*) Fill the large jar with electrolyte.
(*ii*) Use the plastic bottle as a syringe. Squeeze it firmly to expel some air, then connect it by a tube to the top of the burette. Open the burette's tap and release the bottle so that it sucks electrolyte up into the burette. Close the tap. Remove the bottle.
(*iii*) Repeat (*ii*) until the burette is full. Switch on the current and adjust it to 1 A. (*Warning:* the position of the burette relative to the electrode has a marked effect upon the current.)

After that adjustment, switch off the current. Bring the water up to the top of the burette again.

Then let the current flow for a measured time, say 20 minutes.

Switch off, and read the volume of the hydrogen.

Taking the density of hydrogen as $8\cdot4 \times 10^{-4}$ g/cm³ (at one atmosphere and room temperature), calculate the mass of hydrogen liberated. From that, calculate CHARGE/MASS, e/M, for hydrogen ions.

Note: It is unnecessary to adjust water levels before reading the volume of hydrogen, to correct for the pressure of the head of water. It would make a difference of a few per cent at most and would add an unscientific complication because this is clearly a fairly rough measurement.

Measurements with hydrogen show that one kilogram of hydrogen is liberated when $9\cdot6 \times 10^7$ coulombs pass across.★

Point out that it seems very likely that the current in water (really, water + acid) is carried by particles of hydrogen, each of them carrying the same size of positive charge.

In Year 4, we suggested a demonstration experiment, to show that the quantity of hydrogen liberated is directly proportional to current and to time, and therefore to the total electric charge carried across. This does not prove that the hydrogen is travelling across as atoms all alike or that these atoms, if they exist, all carry electric charges of the same size; but those are the easiest assumptions to make.

If the carriers (ions) all have the same charge

★ More precisely, $9\cdot65 \times 10^7$ coulombs for $1\cdot008$ kg of hydrogen. This makes e/M $9\cdot57 \times 10^7$ coulombs for $1\cdot000$ kg of hydrogen.

and mass, e/M for a single hydrogen ion must be 9.6×10^7 coulombs/kilogram. We compare that with the value obtained for the electrons in our experiment:

for H ions	for electrons
about 10^8	about 2×10^{11}
coulombs/kg	coulombs/kg

The value for electrons is nearly 2000 times greater. Electrons must have 2000 times smaller mass or 2000 times bigger charge, or some combination of those disproportions.

We cannot give pupils clear experimental evidence that the charges *are* the same in size: but we can assure them that a number of different types of experiment converge to indicate the same size; and Millikan's experiment suggests that all charges on ions are one electron charge or a multiple of it.

We may point out that if a hydrogen ion is made by knocking one electron off a neutral atom, the charges must be equal and opposite.

All this is a mixture of reassurance and plausible assertion, which is well vouched for in our own experience but not in what we can show to pupils.

If the charges *are* the same size, electrons must be very much lighter than atoms. In fact the electron has a mass only 1/2000, or more accurately 1/1840, of the mass of a hydrogen atom. An electron is only a chip off an atom.

At this point we should give the hydrogen ion, the atom which has lost an electron, its modern name, a '*proton*'.

{In the electrolysis of water, pupils can also see gas (oxygen) being liberated at the other electrode; and they may have heard about negative ions travelling in electrolysis as well as positive ions, though in the opposite direction. So they may object to our saying that *all* the current is carried by positive hydrogen ions alone. Unfortunately, this is both true and untrue; and the detailed story would divert attention from the essential discussion here. In the middle region between the electrodes, the current is carried by positive and negative ions moving in opposite directions with different speeds. But very near an electrode ions of one kind are driven away when electrolysis starts; the current is then carried almost wholly by ions of the other kind, which therefore have to move faster in that region, just before they arrive at the electrode.}

MASS OF ELECTRON, MASS OF PROTON, AND THE AVOGADRO NUMBER

Since pupils have heard about Millikan's experiment in Year 4, and have been told the result, $e = 1.60 \times 10^{-19}$ coulomb, they can now calculate the mass of a single electron:

$$\text{mass of electron } m = \frac{m}{e} \times e = \frac{e}{e/m}$$

$$= \frac{1.6 \times 10^{-19} \text{ coulomb}}{1.8 \times 10^{11} \text{ coulomb/kg}}$$

$$= 9 \times 10^{-31} \text{ kilogram}$$

We can also work out the proton's mass from e and the electrolysis of water, though that may well have been done in Year 4, as soon as Millikan's experiment was done.

Mass of proton From Millikan's experiment and electrolysis,

$$M = \frac{1.6 \times 10^{-19} \text{ coulomb}}{96 \times 10^6 \text{ coulomb/kg}}$$

$$= 1.67 \times 10^{-27} \text{ kilogram}$$

POSITIVE IONS

When an atom (or molecule) has lost an electron, the remainder is positively charged and has most of the original mass. We call that remainder an *ion* (a 'traveller in an electric field'). We can analyse any mixture, atom by atom, by making positive ions from it then using electric and magnetic fields to accelerate the ions and focus their paths, much as in the fine-beam tube measurement for electrons; but with the fields reversed (for positive charges) and made much stronger (for the large mass of an atom).

With electrons we find a single value of e/m whatever the source. But for positive ions e/M is different for different elements, and quite often the same element has several values (isotopes). Then the record is spread out so that it looks like a line spectrum; and the instrument producing it is called a *mass spectrometer*.

MASS SPECTROMETER

First make ions by bombarding the sample with electrons from an electron gun. Then accelerate the ions in an 'ion gun'. Then use a magnetic field to sort them out by values of e/M and focus their streams to make sharp marks on a photographic film.

The diagram shows a simple modern mass spectrometer (Dempster model). A small sample of gas is fed into the 'ion gun' region of the

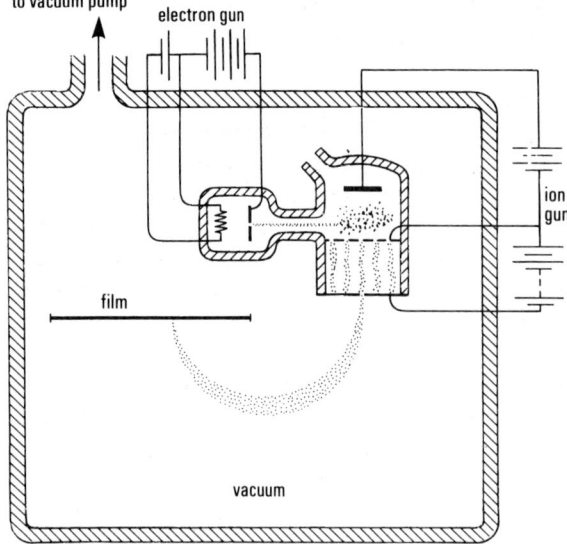

to vacuum pump

electron gun

ion gun

film

vacuum

All ions which have the same mass are brought to a focus by the magnetic field, which is perpendicular to the paper.

apparatus and excess gas is continually pumped away. The sample is bombarded by electrons from an electron gun at one side, so that ions of the sample gas are manufactured.

A weak electric field is applied to the region where the ions are made. This field drives them gently through a grid. They arrive beyond that grid with very little energy. *After that*, they are accelerated through a much larger voltage.* So they emerge from the muzzle of the ion gun with kinetic energy which is essentially given by that main gun voltage.

Since all the ions emerge with the same kinetic energy all with the same mass emerge with the same momentum. So all ions of any one mass will follow the same circular orbit, held by the strong uniform magnetic field perpendicular to the stream.

Even with a fine hole or slit at the muzzle, the emerging stream splays out through a small angle, but the circular orbits of the ions will focus sharply after a half circle. A photographic film (or a

* Like the method used by a tactful restaurant manager to remove an undesirable customer: the customer is led gently to the door, then outside he is given a violent push.

collector for an amplifier and electrometer) will record a focused stream of ions of each mass at an appropriate place in that region.

At this stage we should not bother pupils with the refinement of focusing streams of ions – though that has always been a very important problem in designing machines from earliest mass spectrometers to modern accelerators.

MASSES OF ATOMS: ISOTOPES

We should tell pupils something about the results yielded by mass spectrometers.* *Pupils' Text* includes a discussion of the evidence for isotopes pointing out that the occurrence of many elements with near-whole-number atomic masses suggests the existence of a building block of mass 1. It cites the strange case of chlorine with an atomic mass of 35·5 and the unravelling of the mystery with the development of the mass spectrometer and its explanation in terms of isotopes.

A mass spectrometer gives e/M for carbon ions, $^{12}C^+$, twelve times smaller than for hydrogen ions, and for oxygen ions, $^{16}O^+$, sixteen times smaller. One might have expected e/M for chlorine ions, Cl^-, to be about $35\frac{1}{2}$ times smaller than for H^+. But no such value appears. Instead, there are two orbits made by streams of ions in the magnetic field: one corresponds to a value of e/m, 35 times smaller than e/M for hydrogen ions; the other, 37 times smaller.

This tells us clearly that chlorine atoms come in two masses, a lighter kind with atomic mass 35, and a heavier kind with atomic mass 37. But the '35' stream is nearly $3\frac{1}{2}$ times as plentiful as the '37' stream; that is why together they average about $35\frac{1}{2}$.

ATOMIC MODELS

Tell pupils that we always find the same value of e/m for electrons whatever source we use – hot filaments of one metal or another; photoelectric effect; bombardment of gases by more electrons; enormous electric fields tearing electrons out of cold metal; and even radioactive nuclei emitting beta particles.

* In the course of the nineteenth century, scientists made chemical measurements of atomic masses of all elements then known, on a scale that took 1·0000 for the mass of the hydrogen atom. Nowadays the scale has been changed by international agreement to make the number exactly 12·0000 for the common carbon isotope ^{12}c.

(This is not the moment to mention the smaller values of e/m that we obtain when we observe electrons moving at very high speeds. That can come later as a welcome modification when it will not disturb the main story. There is every assurance that the electrons which have abnormally high mass when we see them moving very fast return to normal when we have slowed them down relative to the observer.)

So we think of electrons as common universal ingredients of matter, tiny chips of atoms, all alike, all with the same mass and the same negative charge.

Positive ions seem to be the 'rest of the atom', carrying most of its mass and having, therefore, different masses for atoms of different chemical elements.

Mass-spectrometer records of positive ions show a great array of different marks, for atoms of different elements and for isotopes of the same element; but electrons make a *single* mark: they are universally identical.

So, pupils may *at this stage* picture an atom as a round blob out of which an electron can be chipped. Therefore since matter is normally electrically neutral, the rest of the blob is positive – we could not tell whether it is a diffuse body of positive electricity like a pudding or made of small knobs of positive electricity.

However, we can knock more than one electron out of an atom. Analysis of positive ions made by bombarding gases with electrons shows that some ions have twice, three times, etc., the normal e/M, suggesting that they have multiples of the basic electron charge. Very early experiments showed that oxygen ions can have several electron charges and mercury ions as many as eight positive charges. Therefore such atoms contain several electrons.

So three-quarters of a century ago, scientists pictured atoms as a sort of pudding of positive electricity with negative electrons as plums in it. This picture was not necessarily intended to be a description of reality but just a way of remembering how atoms behave under electrical attack.

Presently, when pupils look at the fantastic scheme of spheres within spheres imagined by Greek geometers to describe the motions of the planets, they may laugh at the silly ideas of medieval philosophers, who thought that those 'crystal spheres' were so real that a comet passing through must smash them. If pupils laugh, we should laugh with them at those later philosophers who tangled reality with their didactic models; but we should remind them that the Greeks who conceived such 'theories' were exceedingly able, imaginative thinkers who knew quite well that they were describing effective machinery – a scheme that could describe and predict successfully – but not an impossible reality. *The same warning should be applied to atomic models today.*

Further models: nuclei? When pupils ask 'But what about the nucleus?' we should say clearly that nothing seen in experiments described so far conflicts with our picture of an atom as a pudding. As good scientists, we shall not build further details into our picture – such as the idea of a small massive nucleus – until new evidence forces us to do so.

New theory by necessity This is a very important aspect of science which we should teach all our pupils, non-scientists and scientists alike: that the great advances of theory, as in our pictures of atoms, are not just made by imaginative flights of fancy – the scientist's paintbrush twirled at random – but are usually forced upon us by the growth of surrounding knowledge. True, our models always contain an imaginative element; but we try now (as scientists have tried for the last 300 years) to avoid unnecessary imaginative frills (Occam's razor; and Newton's '*hypotheses non fingo*' – 'I do not feign hypotheses').

Young people would like the frills; they would like to think of electrons crawling over metal surfaces like beetles – they would almost let us tell them how many legs those beetles have! They would like to think of electrons in atoms whirling round a nucleus on sharply cut elliptical orbits. Young nuclear enthusiasts might like to say a neutron *contains* a proton and an electron inside it. They are not pleased if we express doubts and ask whether a 5p coin contains two 2p coins and a 1p coin rattling round inside. They do not welcome our scientific caution. In setting forth that caution, we should make it clear that we thereby aim at greater wisdom and fuller knowledge and are not just expressing an insecure agnosticism.

Notes and comments on the e/m experiment

{**The need for two deflecting fields** In making measurements on electron streams (or streams of other charged particles, such as positive ions), there are two quantities we do not know: e/m, and v, the speed of the particles; so we need two separate measurements. Any measurement of the effect of an *electric* field yields e/mv^2. Any measurement of the effect of a *magnetic* field yields e/mv. If we make each of those two kinds of measurement, we can extract e/m and v; but it is no good making two measurements of one kind instead.}

{For example, if we measure the *gun voltage* used to accelerate the electrons (which gives them kinetic energy) and measure the deflection of the stream by an *electric* field, we have two measurements of the same kind; and each will tell us only e/mv^2!}

{**Early history** The reason why early experimenters, such as J. J. Thomson, used electric-field deflections instead of the gun voltage was because they could not command streams of particles which all had the same kinetic energy; their particles were manufactured in the plasma-like mess in a discharge tube and ranged in energy from the full applied voltage downwards. Even that applied voltage was often uncertain in value. No wonder J. J. Thomson used electric-field deflections and had to look at the sharp edge made by particles with maximum energy.}

{No wonder he spread positive rays of varying energies over a parabola in his brilliant experiment that proved the existence of isotopes and led to modern mass spectrometers.}

{However, those are now historical methods, dating back more than three-quarters of a century in the rapid history of atomic physics. With our pupils, we should not treat those methods as part of modern knowledge but should relegate them to special studies in the history of science. Moreover, they were conducted by master physicists to whom the geometry of electric field patterns and fringe effects were child's play, in contrast with our pupils for whom every simplification of mathematics adds greatly to the chances of understanding.}

{Even for A-Level physics specialists, some critics wonder whether the historical methods deserve attention when newer methods and further knowledge are pressing for inclusion. If we could listen to colleagues in another science discussing great experiments in their history, such as a brilliant investigation of a mixture of nitric oxide and air – in an age when even names of gases were confusing – we may share the doubts of those colleagues about the use of historical teaching *for beginners*, and we should turn those doubts upon the teaching of our own modern science.}

{**A null method** The early experimenters had one useful trick. While one of their measurements was made with a single applied field (usually magnetic) the other measurement was made with electric and magnetic fields, applied simultaneously and adjusted to produce no deflection to yield an estimate of v. This might seem to offer a very simple way of making measurements ourselves with a hot cathode tube containing large plates for electric field deflections. We could make the forces exerted by the two fields equal and opposite; and we could state that as an equation without having to measure any deflection. But, for any simple use of this method, the two fields must be coterminous – they must extend over the same region of the path of the stream. We cannot secure this in any available apparatus. So we could only use this method for very rough measurements. And, having obtained v by this null method, we should still have to measure a deflection to obtain e/m as well.}

{**A direct measurement of speed** Instead of making two measurements (one with electric field, one with magnetic field), we could make just one, and do a separate direct experiment to measure v. The latter is done by some form of 'chopper': the

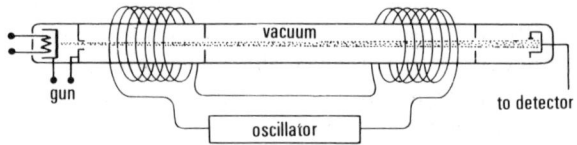

stream passes through slits in two screens far apart, with synchronised 'valves' to interrupt the stream just before each of the slits. The 'valves' are usually electromagnets carrying rapidly alternating current, whose field swings the stream off the slit, so that it can only pass through the slits in periodic bursts. If the succession of bursts gets through both slits we know the time of travel between one slit and the next in terms of the frequency of the oscillating deflecting field. Although this chopper method has been used for electrons (and for positive ions) it is much too difficult for a teaching demonstration.}

{**A rough estimate? Measuring the orbit** The electron stream is seen as a faintly glowing circle in a large glass bulb, in a half dark room. How can the teacher, let alone the pupils, measure the orbit diameter quickly and easily? By holding a ruler up in front and making *a guess* at the orbit diameter? That will yield rough estimates of v and e/m, and, in our exploration of the micro-world of atoms, rough estimates are good science. We might expect them to be correct within 10 per cent. Such a guess at $2R$ should be correct within 10 per cent – certainly within 20 per cent.}

{**Improving precision?** The fine-beam tube has been used for teaching in many laboratories. A careful measurement of orbit diameter is difficult; and it has a complicated history of attempts to make it easier. Teachers have devised skilful schemes for measuring it with some precision – illuminated scales, double scales to avoid parallax, special callipers, observing telescopes, etc. These methods do facilitate more precise measurements, but they also make the experiment more complicated and difficult for pupils to see and remember. Even when handled by a teacher who has had considerable practice, any such special device adds weight to an experiment that is already almost too heavy for many a pupil.}

{**Seeing and doing** Furthermore, if we emphasize measures to improve accuracy we miss the point of the experiment: at this stage it is to bring pupils into contact with a real measurement of electrons. The importance lies in their seeing and doing, in the principle of the experiment and its general success. Accurate measurements can come at other times and places.}

{For average pupils, a successful dive into the micro-world to make a real measurement should be a great achievement. Emphasis on accuracy might make the whole business too hard, or, just as bad, it might turn the experiment into an anxious game to 'get the right answer'.}

{With very able pupils we should emphasize the idea that it is good science to make a rough estimate first. Then if those pupils want to devise refinements for accurate measurements, well and good: they should repeat the experiment as they wish. (Even then, we hope teachers will point out the contrast between making a fine estimate of a fundamental quantity and trying to get the right answer.)}

A note about pupils' demonstrations for revision

Whatever the changes in the examination pattern ahead, most pupils in our Year 5 will be expecting something like a Nuffield Physics O-Level Examination at the end of this Year, a test of their knowledge and understanding with aims that the Examiners keep in mind, not only in framing the questions but also in marking the answers.

Pupils will be concerned about revising for such examinations and teachers will be anxious to give help and make candidates feel secure. Yet intensive revision of notes and learning of formal statements would not be as helpful for a Nuffield examination as for some other tests which rely more on mechanical memorizing. At the extreme, where understanding is looked for, it is doubtful whether intensive coaching can instil understanding or even help pupils to show what they already have of it.

So, while we leave the choice of revision, in quantity and nature, entirely in the hands of each teacher, we suggest that formal revision will be less profitable than one expects – except for building a sense of security – and that it probably deserves much less time than usual. On the other hand refreshing contact with their own experimenting is likely to give pupils valuable help both in knowledge and in spirit.

Therefore we suggest that a good deal of laboratory time could well be given to pupils' demonstrations for revision – a circus in which pupil groups give demonstrations to each other. In the Appendix we put a list of suggested experiments, some very easy, yet fun to do; some giving personal contact with the apparatus; and some practically new projects: but we trust that teachers will make their own lists.

ASTRONOMY: Three chapters for pupils' reading

Our study of planetary astronomy is intended to give pupils a full example of the development of physical theory. Since this involves sketches and discussions rather than pupils' experimenting, we provide it as a set of chapters in *Pupils' Text* for pupils to study through their own reading.

If pupils read all of these chapters for homework, with only an occasional demonstration or other help from their teacher, it will leave the teacher free to continue with discussions and experiments in other topics in class.

This study of astronomy is a large, important part of the Year's activities; and Nuffield Physics O-Level examiners would certainly expect to ask questions on it. But, in framing questions and in marking the answers, they would certainly remember that the topic is now offered for pupils' reading.

OUTLINE OF TOPICS IN ASTRONOMY

Chapter 3
Thinking in science; facts and early fancies

What we see: a short account of the observed motions of stars, Sun, Moon, and planets, to show what is to be 'explained'.

Schemes to explain and predict: Even primitive attempts to make Nature seem reasonable and consistent, bound by some rules, are the beginnings of theory. The Greek raised 'reasonable' to a new level by their machinery with constancies: constant radii and constant speeds. Theirs was strong theory already – fruitful as well as re-assuring.

Early measurements: Hellenistic astronomers estimated the radius of the Earth (probably fairly accurately); the size and distance of the Moon, and the size and distance of the Sun (with an enormous error). Such measurements gave the theoretical machinery a new aspect.

Chapter 4
New Developments

New simplicity: Copernicus put forward a simpler view of a solar system, with telling arguments in favour.

Higher fidelity: Brahe, the great observer with a passion for accuracy, gathered and stored the knowledge which enabled Kepler, 'the law-giver of the heavens' to extract laws of planetary orbits and motions.

Convincing publicity: Galileo taught Copernican theory and argued compellingly in its favour. With his telescope he saw Jupiter and its moons and argued that they constituted a real model of the solar system.

Chapter 5
The Grand Theory: Universal Gravitation

Newton, brooding on problems of the solar system; building on Galileo's mechanics; taking hints from Kepler's work; inventing calculus, and guessing at a law of gravitation, built a magnificent theory: fruitful, economical, and widespread.

Although we have been aiming at Newton's work, pupils need to read the earlier history so that, rather than meet a grand theory ready made, they see theory developing as men are forced to take new views.

EXPERIMENTS AND NOTES FOR TEACHERS

The reading will lose much of its force if its content is also taught in class. So we shall not repeat the material in this *Guide*; but we hope teachers will look at it in *Pupils' Text* for their own interest.

Notes

From time to time we have comments to offer to teachers and we give them in Chapters 3, 4, and 5 which follow.

Experiments

There are a few experiments that pupils should try for themselves – such as sketching an epicycloid freehand, and drawing an ellipse with a loop of thread – and these are described in *Pupils' Text*.

There are also some demonstrations which will help the reading considerably. We give details of those in the following pages.

In this astronomy section of the *Pupils' Text*, both demonstrations and experiments are mentioned less formally than usual, so as not to interrupt the flow of reading. Therefore their listing in this *Guide* is more important than usual. Pupils will need encouragement to try the experiments as far as possible, and they will need to see the demonstrations, or the material will not come fully alive.

We suggest that teachers consider the advisability of designating one session in each of the weeks devoted to this part of the programme to the discussion of difficulties which have been encountered in the reading and to the demonstrations and experiments described in this Guide. We should remember that thinking in three dimensions is a skill requiring practice and we should not be surprised if some pupils meet difficulties in understanding in this area. Such difficulties can readily be cleared by reference to a simple three-dimensional model of the sort described on p. 48.

CHAPTER 3
Thinking in science:
Facts and early fancies

NOTES

A reason and a plea Before pupils start reading, give them the reason for studying astronomy – describe the aim of this part of the programme. Explain that they are not going to learn astronomy for the sake of its factual knowledge, but in order to follow the ingenious thinking that Man has done to explain the events in the sky, as well as to predict some of them.

Warn them that the reading will ask first for a special trick: that they should forget their modern knowledge of a spinning Earth moving round the Sun and *imagine* they are back with early ancestors. To see how knowledge developed, they are asked to start with things the earliest astronomers saw and the stories they told.

Teachers may need to assure pupils that such a backward jump with imagination is a clever and rather difficult move, not childish nonsense. Explain that we make that strange beginning not because we want pupils to start learning astronomy all over again, but because we want them to see how scientists arrived at present knowledge.

One might say to pupils: Please pretend you are back with our earliest ancestors. Why did they worry about the Sun, Moon, and stars? What did they find out? How did they find out?

If we ask pupils now how they would prove to their younger brothers and sisters that the Earth is round, they can probably give us some answers. Ask how they would prove that the Earth spins, and we may find that they put their trust in a 'special' pendulum experiment.* Ask how they would prove that the Earth goes round the Sun, and pupils will indeed have difficulties – Galileo could only assert it. With some class groups, starting up a discussion like this makes a good beginning; with others it only produces a sense of frustration. Each teacher must judge.

Factual knowledge Pupils will need to know some facts of the observed motions of the Sun, Moon, stars, planets (and, later on, comets). They may have learned all that at an earlier stage of school life. But most of our pupils live an urban life and even those who live in the country may have only occasional chances of seeing a clear, starry sky. So, unlike earlier generations, and people in countries with cloudless skies, our pupils need to be told some 'obvious' things – and need, if possible, to see some of those things.

Help needed: the zodiac All the stars revolve together. Only the Sun, Moon, and planets have somewhat different motions: they swing round with the star pattern's daily westward motion, but they also crawl slowly eastwards through the pattern along a slanting path. The exciting things are the motions of Sun, Moon, and planets, and the exciting part of the motion of each of these is its strange eastward lagging or wandering through the star pattern rather than its rapid daily motion. So, astronomers at a very early stage of the science started leaving out that daily motion – 'subtracting' it. In other words they imagined the daily motion stopped, or 'frozen', and then catalogued the remaining motions of Sun, Moon, and planets. That was an enormous step forward, a difficult intellectual jump. Some pupils find this a difficult step to understand. Offer them help.

All the slanting paths characteristic of those lagging motions of Sun, Moon, and planets lie in a band called the *zodiac*, with the Sun's yearly path as its central line, the *ecliptic*.

Professional astronomers think of the general eastward sliding of the planets round the zodiac as the main, normal motion, and they call the reversed motion in the loops which the planets appear to make in each of their years 'retrograde'. Here we call:

the *daily* motion of stars, Sun, Moon, planets, *westwards*.

* Foucault's pendulum is a risky experiment, anyway. Quite apart from the Earth's motion, a pendulum whose bob moves in a narrow ellipse will precess – because the period of the component along the major axis is slightly greater. So unless such a pendulum is released very carefully by burning a thread to swing with linear motion, it may precess, *either way*, for a different reason.

the *yearly* motion of the Sun round the ecliptic, *eastwards*.

the *monthly* motion of the Moon round its orbit, *eastwards*.

the *general* motion of each planet round the Zodiac, *eastwards*.

the '*retrograde*' motion of a planet in each loop, *westwards*.

The word 'retrograde' is not used in *Pupils' Text*.

The 'year' of a planet's motion The periods of orbital motion of the planets guided the Greeks in placing them in an order, and were used by Kepler in discovering his third Law; then they played essential parts in Newton's theory.

A planet's period – the time it takes to get round its orbit – depends somewhat on our viewpoint. The value used in *Pupils' Text* is the 'true' period or 'planetary year' as an observer on the Sun would see it.

As we now picture the solar system, we might draw a radius from the Sun to a planet and on out to the stars, to mark its position in its orbit. As that radius turns through 360° the planet goes once round the orbit and returns to the same place, as it would be seen by an observer on the Sun, against the star pattern.

However, during that 'true' period of the planet's motion round its orbit, the Earth moves to a different position, and an observer on the Earth would not see the planet back at the same place 'among the stars'. The planet's *apparent* period, recorded by an observer on the Earth, is a modified compound value derived from the planet's 'true' orbital motion and the Earth's orbital motion. If we were giving a proper account of the planetary picture seen by early observers, we should give the 'apparent' periods. Then we should have to disentangle the 'true' periods from them when we came to the Copernican picture of the solar system.

In our present teaching we want to give a simple, clear picture of the problems that led to theory rather than explore such special details. So, we give only the 'true' periods, in *Pupils' Text*.

Eclipses Eclipses deserve a brief mention, and explanation, but not with the usual diagrams of umbra and penumbra. Those diagrams give names to be learnt without being sufficiently true to scale to give good knowledge.

How many of us realize that the full shadow of the Moon is a cone of angle only $\frac{1}{2}°$, whose very tip only just reaches the Earth? How many realize that the shadow of the Earth, which must itself have narrowed by one Moon-diameter out at the distance of the Moon, just covers $2\frac{1}{2}$ Moon-diameters as the Moon passes through the shadow in an eclipse? From that, and the $\frac{1}{2}°$ angle subtended by the Moon, we can at once show that the Moon must be about 60 Earth-radii away, some 380000 kilometres – a distance now known with great precision, better than to the nearest metre. See the diagrams in *Pupils' Text*.

Precession of the equinoxes At some time in the factual story pupils must learn of the precession of the equinoxes. As described by early astronomers, from an Earth-centred point of view, it was a very obscure creeping motion of the whole system of stars around a special axis (the axis of the ecliptic). In that form, it is too difficult for pupils to understand. As described by Copernicus, it is much simpler and his description is given in *Pupils' Text*, so that Newton's 'explanation' may be enjoyed.

Astrology, superstition, and our duty in teaching As astronomical knowledge developed, there grew up a body of superstitious belief in astrology. That belief, still alive today, provided a driving force for careful observing of the planets – and a source of finance for astronomers for many centuries.

We might ask pupils, 'What is superstition? Can you explain in a short sentence what that word means?' There seems to be no compact answer to that question, except some dogmatic statements that lead to hot arguments. Some teachers find that this question and its discussion help a discussion of theory. Others consider it an unfortunate diversion. One should be guided by taste.

Even today many educated people look to astrology for some guidance, if only with half-hearted belief. Why should non-scientists prefer a physicist's Newtonian treatment of the planetary system to the more romantic view embodied in astrology? If, as scientists, we believe the preference should go towards Newtonian theory, this throws light on our present duty.

The beauty of Greek theories In *Pupils' Text* we describe the history of astronomy as a basis for

Newtonian theory. But, on the way, we show some samples of Greek theories. Those theories were works of genius that need careful study if one wants to master the details of the machinery and appreciate their success.

But, sir ... However, teachers may meet a difficulty: pupils object, at the Greek stage of their reading: 'But sir, we know this isn't true. We know that the Earth ...' Pupils have been told, much earlier, some things about the Earth and the heavens. They know that the Earth moves and not the Sun. They know quite well that the Earth spins and travels round the Sun in a year. Many of them also know that the planets go round the Sun in a circular orbit, and do not swoop back and forth in loops as they travel round the zodiac.

So the explanation of our plans given at the outset in *Pupils' Text* may have to be repeated – that this is a tricky problem in thinking and understanding, to see how people could build a picture that would fit the facts, and then go further to link the facts with other knowledge, then even further to predict new things to look for.

The main target One might already say to pupils, 'Keep your eye on the main target: *Newton's theory and its rich variety of explanations.* The sooner you get there the better, as long as you have seen enough of the earlier stages on the way to be ready to welcome a general theory.'

Ingenious models: misleading here

In *Pupils' Text* we show Greek schemes for stars and planets, and then Copernicus' solar system on the way to our target: Newtonian theory. One is tempted to make mechanical models to illustrate the schemes but showing complex models is likely to take too much time and to divert the attention of the pupils from the main advance to the target. We offer the following comments on mechanical models:

1. In this programme which is concerned with the development of a theory, mechanical models of early schemes are not necessary. Nor are elaborate celestial spheres or models of the solar system. Experience has shown that they divert attention from ideas to machinery, from intellectual grasp to interest in mechanical ingenuity.

2. We advise schools *not* to buy an orrery or any other elaborate model.

3. Where a laboratory already has models, they might profitably be used, *if* they can be introduced *lightly* and shown very briefly. A sphere with an internal lamp and pinpricks for stars is *not* a model of a Greek scheme; like a planetarium, it illustrates what we observe rather than showing a Greek scheme.

4. However, where a teacher has devised his own model, we would *not* discourage him. But we would offer two comments: (a) in making the great profusion of models in the past, inventors have found that devices which involve spins about several axes have to be more sophisticated than one would expect; (b) a 'partial' model, such as an open umbrella, helps the teaching quickly onward.

Nevertheless, we do suggest that teachers should show two very simple models to act as catalysts: one improvised from a glass flask, the other from an umbrella.

The simple flask model

Apparatus

1 large 2-litre bolt-head flask   item 87
1 rubber bung to fit flask
1 long knitting needle
1 $2\frac{1}{2}$-cm polystyrene ball
1 tripod with round top, or a ring on a retort stand or, alternatively, 2 retort stands with bosses and clamps   503–6

Preparation

Mount the small ball on the point of the knitting needle, which has been pushed through the bung, so that the sphere is at the centre of the flask. Half fill the flask with coloured water, so that when the flask is inverted, the water level passes through its centre and the ball is half immersed.

Stick a small coloured disk of paper on the round surface of the flask to mark the end of the axis through the neck of the flask and the ball. This also represents the Pole star.

Mark the celestial equator (perpendicular to that axis) with a ring of tape, or a line drawn with a glass-marking pencil.

Then place the flask with its neck pointing downward in a tripod with a round top, or in a horizontal ring on a retort stand, or, failing that, support it with a clamp and a second boss to prevent the clamp turning under the weight of the model.

a. Celestial sphere

Procedure

Tilt the flask so that the polar axis is inclined at about 50° to the horizontal. The water then provides a horizon and the small ball represents the Earth at the centre of the celestial sphere.

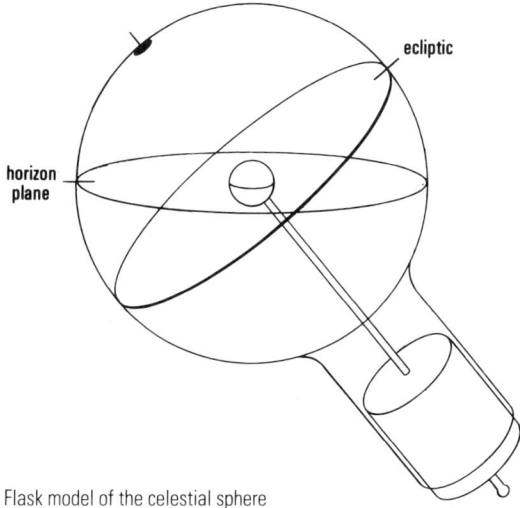

Flask model of the celestial sphere

Add a band of coloured tape to represent the zodiac which includes the Sun's ecliptic path. This band should be at $23\frac{1}{2}°$ to the equatorial plane of the model.

Hold the neck of the flask and turn it about its axis. This imitates the daily motion of the stars. But it does not show the lagging and wandering motion of the Sun, Moon, and planets. These are best explained when the daily rotation can be ignored.

b. Precession of the equinoxes

Mark a new axis on the flask perpendicular to the ecliptic: stick a small disk of coloured paper where the axis meets the glass. An ideal arrangement is to fix a small suction cap there and another at the pole.

Hold the model so that it can revolve very slowly about the new axis. The pupils must imagine that the whole model is also spinning very rapidly (10 million times faster) round the polar axis.

At this stage, it will help if twelve small orange disks are stuck on the ecliptic band to represent the position of the Sun at monthly intervals.

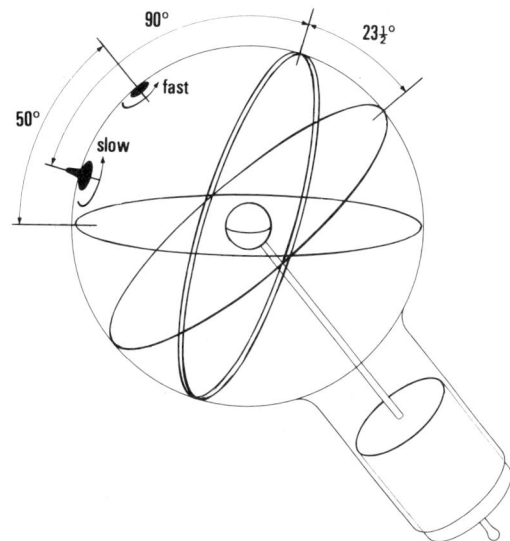

Flask model to show precession of the equinoxes

The crystal spheres When they read of invisible spheres, some pupils may say the schemes are stupid. We hope teachers will argue in their defence.

Even the early scheme of spheres was far from childish or stupid; it was a brave attempt at 'explaining' the heavens and giving reassurance to ordinary thinking people – as well as helping to make crude predictions. The early forms were over-simple, in that they failed to show the retrograde loops in a planet's motion, and failed to show the irregularities in the Sun's motion from season to season.

A scheme of spheres was a tale saying, 'It is all *reasonable*; there is machinery which carries the stars and planets round; it fits together and runs with simple rules; there is nothing to fear.' But the scheme gave no hint of the way in which the motions were started or maintained.

Later schemes of many concentric spheres and then of circles with eccentric arms and sub-circles were devised to fit the facts more and more closely, but always with insistent constancies.

Constancies: an essence of scientific description The Greeks insisted on spheres for their machinery and made those spheres revolve at constant speed. These were not whimsical assumptions made for artistic delight. Some such assumptions are *essential* for a scientific description, which is what the Greeks were aiming at.

When we want to describe some behaviour in nature in the compact way that scientists like, we have to extract some constancies. Each scientific law that we state (usually derived from experiment) is really a statement about something that remains constant, independent of some other changes or details.

Thus, in building science, we try to single out things that are constant. PRESSURE times VOLUME is constant in Boyle's Law. STRESS/STRAIN is constant in Hooke's Law.

If we could not make use of such constancies in our descriptions, we should go mad with the profusion of irregular details. One might almost claim that every natural law can be stated with the word 'constant' in it somehow.

Greek astronomers had to express their knowledge of heavenly motions in statements that contain some constant elements – otherwise they might just as well have ascribed the motions to wayward gods or demons. A sphere has *constant radius*, the same in all directions; this gave it a great advantage in the Greek view, as part of the machinery. And each sphere was a given a *constant speed* of spinning – again, without that constancy, the description would hardly make nature seem reasonable or easy to understand. As the descriptions were elaborated to fit the facts more closely, the Greeks would add more spheres within spheres, each with its own constant motion, rather than lose that essential characteristic. Later, when the profusion of spheres itself lost the attractiveness of simplicity, Greek astronomers modified their insistence on constant speed, but *they installed other constancies instead.*

Simplicity in theory To make a good theory in science, we must have basic principles or assumptions that are simple; and we must be able to derive from them a scheme that fits the facts reasonably closely. Both the usefulness of a theory and our aesthetic delight in it depend on the simplicity of the principles as well as on the close fitting to facts. We also expect fruitfulness in making predictions, but that often comes with these two virtues of simplicity and accuracy. To the Greek mind, and to many a scientific mind today, a good theory is a simple one that can cover all the facts with precision. Nowadays we also expect a good theory to provide *language* to facilitate interchange and growth of understanding.

Is the theory true? Questions to ask, in judging a good theory, are:

'Is it as simple and economical as possible?'
'Does it fit the facts as fully and closely as possible?'
'Does it fit with other knowledge and theories?'
'Is it fruitful in predictions?'

If we also ask, '*Is it true?*', that does not seem a proper requirement. We could give a remarkably true story of a planet's motion by just reciting its locations from day to day through the last 100 years; our account would be true, but so far from simple, and so spineless, that we should call it just a list, not a theory.

In talking to young would-be scientists today, we urge them not to be satisfied with just collecting specimens, or facts, or formulae, lest they get stuck at a pre-Greek stage.

Eudoxus and Ptolemy constructed their astounding schemes as good intellectual machinery. We should be surprised if they thought their spheres or rotating arms were really there in the sky. It was others who invested the mathematical machinery with solid reality.

Pupils' Text provides a sketch which shows a medieval view of the simpler system of spheres in which the region outside the outermost sphere (of stars) has a suggestion of 'celestial machinery' – the power-house (*primum mobile*) that every Greek scheme would need if its machinery were real. That region also came to be called the empyrean paradise – and to be identified with the Christian heaven, the habitation of God and all the elect. The original astronomers may have imagined the outermost starry sphere as a theoretical device for describing the motion; but it soon took on an air of solid reality. Heaven was clearly beyond that sphere. This picture of heaven became so strongly established that the Copernican view, which did not need a sphere for the stars, but placed them at all kinds of distances in remote space, met with dismay and opposition.

Ptolemy's highly successful scheme Ptolemy's scheme remained in use for a dozen centuries. But to show how much simpler Copernicus' scheme appears – when once accepted – some teachers may like to sketch a Ptolemaic scheme for the Sun and two planets. The sketches I–IV show the component pictures, reduced to one plane for simplicity. When drawn on transparent sheets as overlays they make a composite picture which looks complicated enough. Copernicus' scheme seems so much simpler in contrast.

The Ptolemaic scheme for the Sun, S, and two planets, P and P'. Diagrams I, II, III, and IV may be sketched as separate transparencies for use as overlays in an overhead projector.

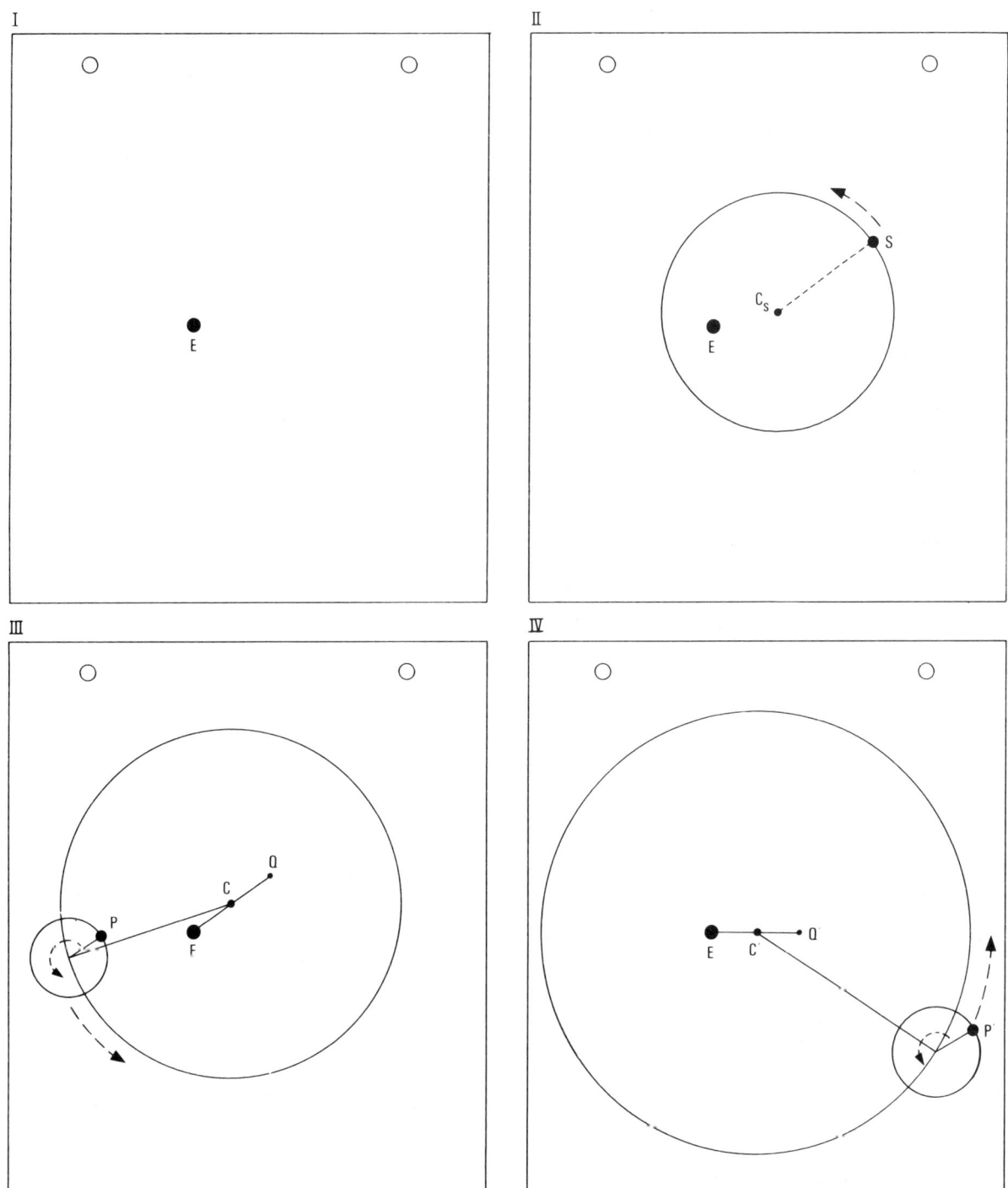

I

II

III

IV

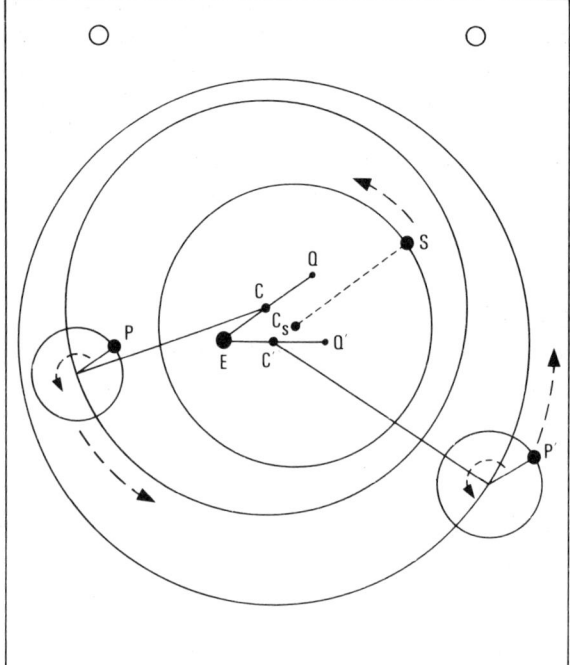

Copernicus himself had to add small sub-circles, but these were to produce small corrections needed because what we now know to be ellipses were being treated as circles. It would miss the point of his simplification to say they made his scheme as complex as Ptolemy's.

The place of Ptolemy Recent research into the work of Ptolemy has cast doubt on the value of his original contribution to astronomical observation and theory. However, the system of epicycles normally attributed to him was the one that most strongly influenced thought about the universe through to the time of Newton and *Pupils' Text* therefore places the customary stress on his contribution. (See Newton R. R., *The crime of Claudius Ptolemy*. Johns Hopkins University Press 1977.)

EXPERIMENTS

Pupils' observations Both urban life and clouded skies allow many pupils to grow up with scrappy, uneven knowledge of events in the sky. They may have felt no need to observe carefully. At this stage, teachers should urge pupils to gain some direct observational evidence by watching carefully.

Pupils' observation 13
Stars in a night sky

Ask pupils to observe the sky on a clear night, at least twice, with an interval of, say, two hours between observations.

Demonstration or Home Experiment 14
Photograph of the night sky

Procedure

Discuss the photo of the night sky in *Pupils' Text*. It was taken by exposing a film in a rigidly fixed camera. If the shutter is held open for several hours, each star makes a large arc of a circle on the picture.

Home Experiment

Encourage any pupils who are interested to make such a photograph themselves, according to these suggestions:

     *   *   *   *   *

Take your camera out of doors on a suitable night and leave it there in its case for about half an hour so that it is at the same temperature as its surroundings. Then, when you open the camera, dew will not form on the lens. Select a large aperture. Fix the camera firmly so that the axis is pointing at the Pole star. Open the shutter to give a time exposure of at least two hours. At the end of that time, close the shutter.

     *   *   *   *   *

The photograph will be more impressive if it includes the silhouette of nearby trees or buildings.

Develop the film in the usual way. Colour film may be used provided the camera has an aperture larger than about f4.

Advise against the use of a telephoto lens. This is a case where a wider angle than normal is preferable.

Pupils' observation 15a
The daily motion of the Moon

Ask pupils to watch the Moon and think about its motion. They should try to locate the Moon with reference to the star pattern. If possible they should follow its motion in the course of a few hours on the same night. They should also note its position at the same hour on consecutive nights.

Note The Moon's diameter subtends about $\frac{1}{2}°$ at the Earth. The Moon moves from west to east through the star pattern 360° in a month, 90° in a week, just over one Moon-diameter in an hour. In that hour, the daily motion carries the star pattern *and* the Moon a much greater distance westwards, about 15° or thirty Moon-diameters.

Pupils' observation 15b
The monthly motion of the Moon

Encourage pupils to extend the previous observations over a month. They should note the position of the Moon at the same hour on each clear night, its position in the star pattern and also relative to the horizon.

Ask (*i*) does the Moon's place in the sky (at the same hour) move east to west from night to night, like the daily motion, or the opposite way?
(*ii*) Can you describe its path through the star pattern?
(*iii*) Where is the Sun at full Moon?

Pupils' observation 15c
The daily and yearly motions of the Sun

Ask them to watch the Sun's apparent motion during a day. Also ask them to note the Sun's position at noon from month to month.

WARNING Remind pupils NEVER to look directly at the Sun, even with 'sunglasses' on.

Ask (*i*) at what time of day is the Sun highest in the sky?
(*ii*) Is its height then always the same?

Pupils' observation 16
Planets

Encourage the pupils to look for the brightest planets – Venus, Jupiter, and possibly Saturn. Venus can be seen in full daylight when favourably placed.

Information on where to look can be found in the monthly articles or maps published in major newspapers, or in annual publications such as *The Sky at Night* (The Times); the *Yearbook of Astronomy* (Sidgwick and Jackson); *Whitaker's Almanack*; the *Astronomical Ephemeris* (HMSO). Some pupils would find *The Observer's Book of Astronomy* (F. Warne and Co.) useful. It is small in size and interesting.

Experiment 17a
Planetary paths: oblique view of an epicycloid

Procedure

Sketch the epicycloid pattern of a planet's apparent path free-hand on the board.

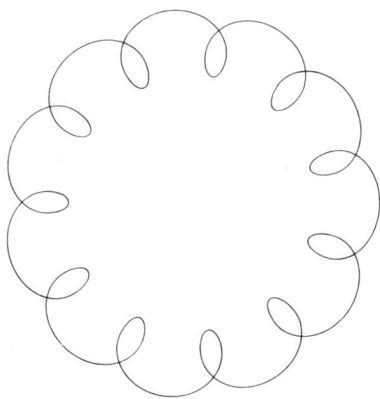

Then ask the pupils to make their own sketch quickly on a piece of rough paper. Alternatively draw a large specimen freely with a felt-tip pen by moving the pen round in a small circle (diameter about 10 cm) while sweeping the hand round a larger circle (diameter 50 to 100 cm).

Tear off a piece of the pattern containing a few loops and hold it obliquely so that the pupils see the pattern almost edge-on. That is how we see a planetary path – since the orbits of all planets, including the Earth's, lie in the narrow zodiac band.

Demonstration 17b
Model of a planetary path

Apparatus

1 turntable	item 154/1
1 small motor on base plate	154/3
1 small ball on rod	
1 dry cell	

Procedure

(*i*) Attach the motor's base plate to the turntable. Arrange a dry cell to drive the motor. Install a small polystyrene ball on the slow-motion wheel of the motor as shown.

(*ii*) *A simpler version.* Hold a small ball high up and

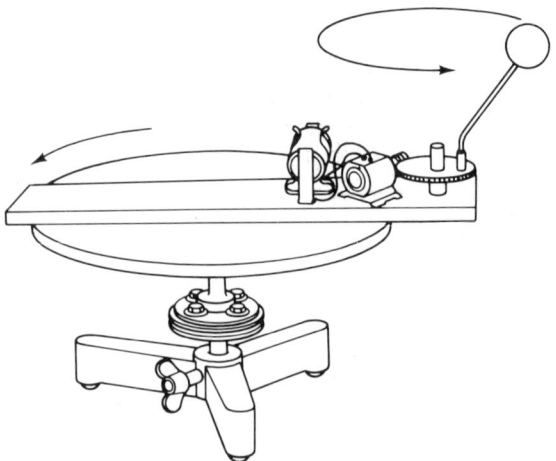

move it round a small horizontal circle. At the same time walk round the room slowly in a large circle.

Demonstration 17c
Planetarium
(*OPTIONAL*)

(*i*) *Visit.* If a visit to a planetarium can be arranged, it will provide a very good model as it will show the motions that are observed.

We hope that teachers will explain that such a visit would make a strong contribution to Nuffield physics teaching.

As a rule, planetarium authorities like to show a sample picture of the heavens for some particular date, with Moon and planets making only small motions during the performance. They hesitate to

run their planetary machinery very fast and show several years' motion in a short time. That necessitates stopping ('freezing out') the daily motion, and it asks for unusual speed. Therefore they hesitate because they know how easily an audience forms misconceptions. However, if the teacher explains to the planetarium authorities the special need to show planetary motions through the star pattern, for an important part of our teaching programme, he would find that such a demonstration could be arranged.

(*ii*) *Home-made model* A simple 'toy' planetarium can be improvised using a two-litre round-bottomed flask (bolt-head type with a wide neck).

Support a 12 V, 36 W lamp at the centre of the flask. Coat the surface of the flask with Aquadag. Then scratch holes in this surface to represent the stars of some major constellations.

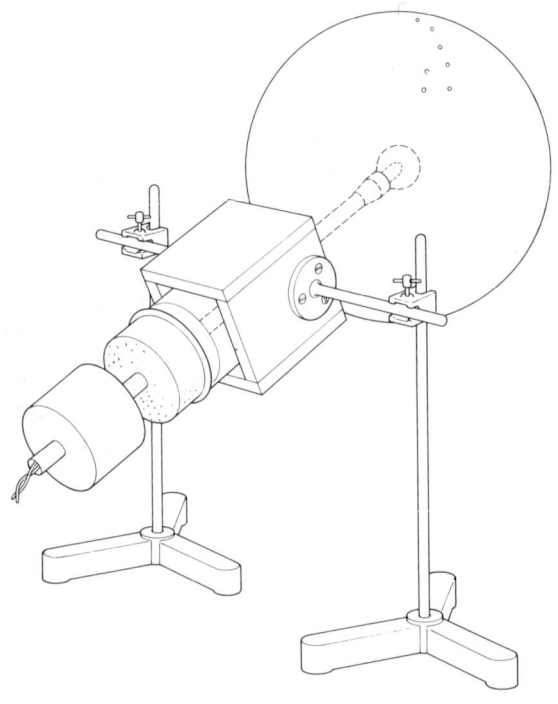

When the lamp is switched on in a darkened room, the spots of light on the ceiling will rotate as the flask is rotated and thus display the simple daily motion.

Note. This is not the time to show an orrery – a mechanical model of the Copernican solar system – even if the school has one. It would represent a jump ahead in our story and it would add confusion. So, although teachers naturally think of an orrery as an easy illustration when they are first talking about planets, we urge them to keep it hidden until they get to Copernicus. Even then we do *not* advise a school to buy one.

Demonstration 18
Simple models improvised from an umbrella

Preparation

When the umbrella is opened, the ferrule represents the Pole star. A few constellations are represented by small paper disks stuck to the underside of the fabric. Chalk marks are not satisfactory – they tend to imprint once the umbrella is closed again.

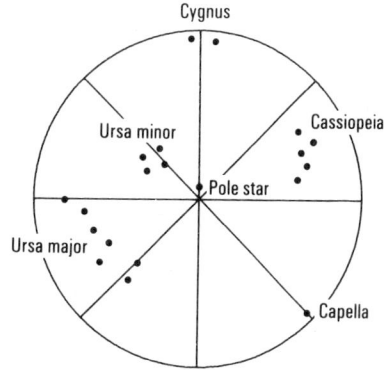

If there are eight ribs, they provide a useful guide in the marking. Pairs enclose 45°, or three hours of time in the daily revolution. The surface can include some circumpolar stars visible from about latitude 45°.

Procedure

a. Celestial sphere

Show the umbrella. Rotate it slowly by the handle in the anti-clockwise direction so that pupils can see the stars on the underside.

b. Model of an early Greek scheme (Thales)

Hold the crook of the handle over a disk (e.g. a saucer) to represent a flat Earth at the centre of the celestial sphere. Rotate the umbrella slowly as before.

Thales

c. Model of Aristarchus' scheme

Represent the Earth by a 5 cm polystyrene ball mounted on a thin knitting needle driven halfway into it. Represent the Sun by a larger ball (e.g. tennis ball).

(*i*) Hold the small ball near the crook of the umbrella, with its rod parallel to the umbrella's handle. First, hold the ball still and spin the umbrella to show what we see.

Model of Aristarchus' scheme

Then, to show Aristarchus' alternative explanation, hold the umbrella still and spin the ball in the opposite direction at the same rate.

(*ii*) Show Aristarchus' alternative explanation of the Sun's apparent yearly motion. Move the Earth ball out a short distance and install a large ball near the crook to represent the Sun. Then carry the spinning Earth ball round the 'Sun' in an orbit, by hand.

This is easier if one dispenses with the stationary umbrella and holds the 'Sun' and 'Earth', one in each hand, or places them on the table. Move the spinning 'Earth' round the 'Sun' with its spin axis always pointing in the same direction.

Demonstration 19
Illustrating parallax

Walk round the room and point out how an observer moving in an orbit sees the changes in the pattern of pupils seated in the class as he moves nearer or further away from various groups. This is done to illustrate the point that it was the absence of noticeable changes in the star pattern which, among other reasons, made Aristarchus' scheme unacceptable to Greek astronomers.

Experiment 20
Model of epicycle system for planets

Apparatus

1 large ball (5 to 10 cm) and 1 small ball (2 to 3 cm)

Procedure

Hold the larger ball at rest as in the sketch to represent a fixed 'Earth'. With the other arm outstretched, hold the small ball and sweep it in a large vertical circle round the 'Earth' as centre. At the same time make the hand of the outstretched

arm revolve quickly round its wrist so that the 'planet' makes a small circle as it moves in its large orbit.

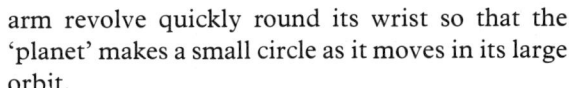

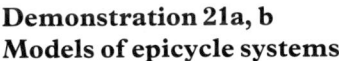

Demonstration 21a, b
Models of epicycle systems

Apparatus

1 turntable with motor assembly   item 154/1 & 3

Procedure

a. Use the small electric motor assembly of Experiment 17b to represent the 'planet' moving in its epicycle. Sweep the whole assembly in a large vertical arc by hand about a fixed 'Earth', while the motor drives the 'planet' in its small circle.
b. Mount the motor assembly on a rotating turntable. Revolve the turntable slowly by hand while the ball on the electric motor revolves faster.

Class experiment 22
Pupils estimate the size of the Earth
(*OPTIONAL*)

Procedure

For this experiment it is necessary for two distant schools to cooperate. The schools should preferably lie almost on a north-south line.

On an agreed day pupils at both schools make observations of the 'height' of the Sun at noon. Each school sets up a pole of known height – say 3 metres. Each pole must be vertical; and the ground around it must be horizontal.

At noon, pupils at each station measure the length of the shadow. They exchange details of apparatus and measurements, preferably by telephone.

Using geometry, the pupils deduce the size of

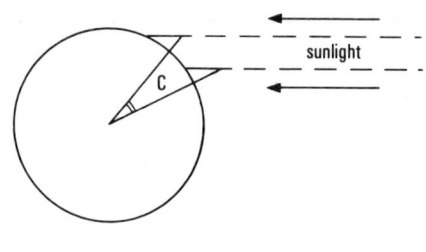

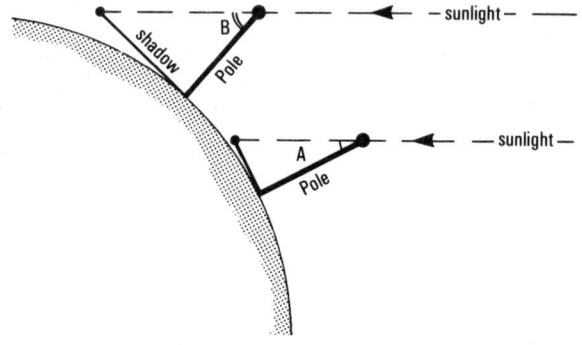

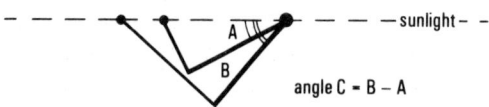

angle C = B – A

the Earth from the measurements. They will also need to know the distance between the two schools (in kilometres) and this can be found from a railway timetable or a road guide – but not from a map, which might seem to beg the question with its latitude lines.

Class Experiment 23
Estimating the ratio of the Moon's distance to its diameter

Procedure

Pupils work in pairs and follow these instructions.

★   ★   ★   ★   ★

Look towards the full Moon and hold a small disk (such as a penny) out in front. Adjust its distance from your eye until it just obscures the Moon. If it needs to be further out than your arm can stretch, a stand and clamp will be needed to hold it – or you might stick it to the window pane.

Ask your partner to measure the distance from your eye to the coin.

★   ★   ★   ★   ★

A penny has diameter 2 cm. So a clamp and stand will be needed if the experiment is done in school.

It will be found that the coin is approximately 110 coin-diameters away. Then the Moon's distance is about 110 Moon-diameters away.

CHAPTER 4
New developments

NOTES

Solar system From Copernicus to Newton the Copernican model grew into full use; slowly at first, partly because Copernicus' book was not easy to read, and partly because traditional thinking had a very strong hold. Remember that no change of star patterns due to parallax had ever been observed. (Small stellar parallax shifts were first observed in the nineteenth century.)

Tycho Brahe favoured a hybrid system in which all the planets except the Earth move in orbits round the Sun, and the Sun carries them as it moves in a circular orbit round a stationary Earth. Tycho had embarked on his lifetime of systematic, meticulous observing with the hope that he would be able to decide between three schemes, the Ptolemaic, the Copernican, and his own hybrid one. He bequeathed his records to Kepler and Kepler is said to have analysed them from all three points of view.

Geometrically, the hybrid scheme must fit all observations as closely as the Copernican one, but to us it must seem unpleasantly unsymmetrical; and probably to Kepler too, when he was trying his five-solid idea and later when he extracted his Law III.

Galileo claimed that Jupiter and its moons showed him a real Copernican system in miniature.

Theory's bootstrap Sometimes theory can lead to measured information concerning its own assumptions. Simple kinetic theory of gases enables macroscopic measurements to yield estimates of molecular speed in gases – estimates that only have meaning in terms of the assumptions that went into the theory. Copernicus' scheme for the solar system enabled him to estimate relative radii of orbits and to make a model to scale. In the Ptolemaic scheme not only were planetary distances indeterminate, but the relative order of the planets was only guessed at from their 'year' periods.

Kepler's search When pupils read about Kepler's work, they may need some sympathetic encouragement – perhaps by a hint of a great use for Kepler's Laws to come just ahead. Remember that pupils are not waiting eagerly for the answer to a great, urgent question, as Kepler was. They do not even feel the strong need to find constancies in Nature, which is one of the main drives in science. Our pupils have grown from the stage of seeing things, collecting facts, of making fuller and fuller acquaintance with Nature; they have watched us extract some rules, codify our acquaintance. Yet we should have been careful not to emphasize those rules in a way that would make them formulae to be learnt by heart for use in answering examination questions. Therefore, rules and relationships are likely to be interesting to pupils, but not yet recognized as the essence of our knowledge.

Galileo's experimenting By reading earlier writers, discussing with piercing logic, and some experimenting, Galileo learnt much about the ways in which Nature behaves. And he taught his knowledge with vigour and conviction – often using a thought-experiment to explore and expound. Though he could work carefully in making instruments, he often experimented very roughly – he was so sure of his understanding of Nature – and would merely quote an experiment to support some exposition.

Galileo and Copernicus' book In one of his dialogues, Galileo describes his first encounter with Copernicus' scheme, when he was quite a young man. He avoided some lectures on it, thinking it would turn out to be a fashionable fad; but then he heard a favourable account from a man who seemed cautious and intelligent; so he proceeded to make a test. Whenever he met someone who agreed with Copernicus he asked him if he had started with a Ptolemaic view. All such people said they had started with Ptolemy, but had been persuaded to change their mind. On the other hand, when he consulted many who still held the Ptolemaic view he found few among them who had even glanced at Copernicus' book, and none who understood it.

He concluded that a book which could persuade readers to make such a serious change was worth investigating. That was probably the beginning of his conversion to a Copernican view, which he later taught with characteristic force.

EXPERIMENTS

Experiment 24
Simple model of Copernicus' explanation of the looped paths of the planets

Procedure

Hold a long, smooth pole of wood, or bamboo, or light metal pipe as a 'sight-line' from the Earth to a planet, say Jupiter or Mars. The pole should be two or more metres long.

Hold one end in the right hand (representing the Earth). Let the pole run loosely through a ring made by finger and thumb of the left hand. Hold the right hand close to the body with the left arm outstretched.

Pupils need to imagine the line of the pole continuing on out to the 'stars' which they imagine to be on the walls and ceiling of the room.

To represent the central Sun, clip a small paper disk to your clothing.

Move the Earth (right fist) quickly in a small vertical circle while moving the planet (left fist) slowly in a large vertical circle. The far end of the pole wags to and fro as well as making general progress across the sky.

Note More elaborate models with an electric torch on the pole have been tried; but they are apt to add confusion rather than clarity. This is a demonstration in which intelligent imagination should play a part.

Demonstration 25
Precession of the equinoxes (Copernicus)

Procedure

Place a large ball (e.g. a football) at the centre of a table to represent the Sun.

The Earth is represented by a small globe mounted on an axis held at $23\frac{1}{2}°$ with the vertical. This may be a small mounted Earth globe or a polystyrene or wooden ball drilled along a radius to a depth of about $\frac{3}{4}$ of its diameter and supported on a knitting needle so that it can spin freely. Bend or tilt that axle to about $23\frac{1}{2}°$ with the vertical.

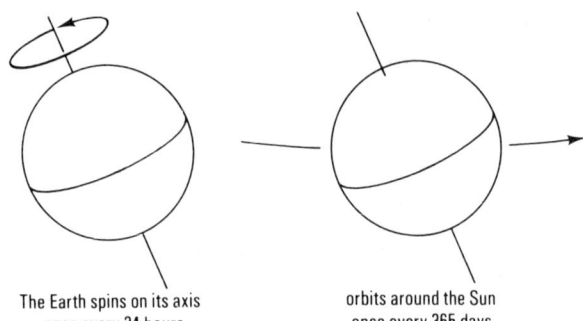

The Earth spins on its axis once every 24 hours

orbits around the Sun once every 365 days

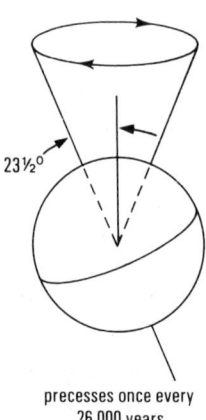

$23\frac{1}{2}°$

precesses once every 26 000 years

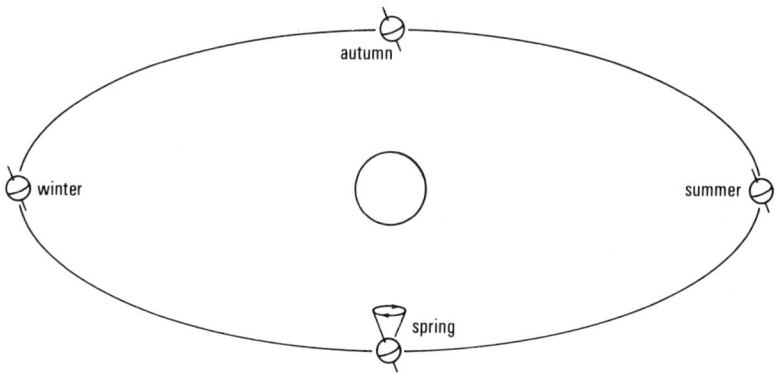

Set the Earth globe spinning and move it round the Sun in a circle on the table with its spin axis always pointing in the same direction. This represents the yearly motion of the Earth. Then show precession by making the spin axis of the Earth globe revolve in a small conical motion about a vertical axis perpendicular to the Earth's yearly orbit.

The real conical motion is a very slow one, with a period of 26 000 years.

Class experiment 26
Drawing ellipses

Apparatus

Each pupil needs :
2 drawing pins
1 loop made from about 25 cm of thread
1 sheet of paper
1 drawing board or sheet of thick card

Procedure

Pupils follow these instructions.

* * * * *

Stick two drawing pins in a board (not pressed far in) about 10 cm apart. Slip the loop of thread over them. Insert a pencil point in the loop and pull the thread taut. Then move the pencil to draw an ellipse, always keeping the thread taut.

* * * * *

Pupils might try drawing ellipses with the pins much closer, also farther apart.

Elliptical orbits Simple demonstrations 'of elliptical orbits are difficult to arrange. Probably the best that can be done are those described here, but neither is very convincing. The orbit precesses, showing the effect of friction.

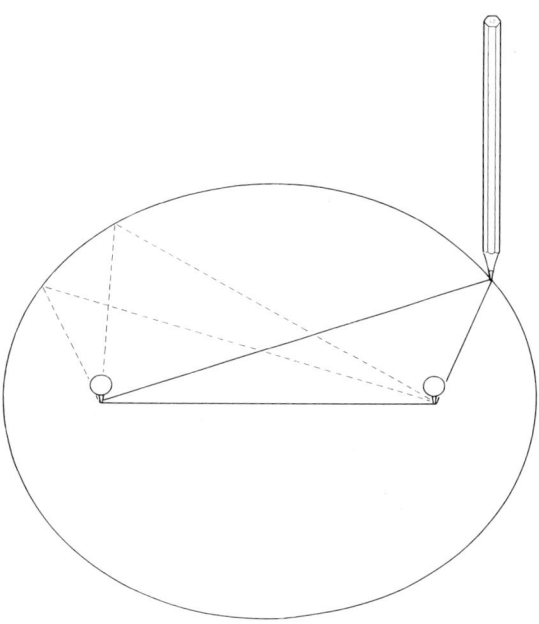

Furthermore, in neither case do the pupils know whether the force is an inverse square one. They are really only qualitative illustrations of oval orbits.

61

Demonstration 27a
Illustration of elliptical orbit (*OPTIONAL*)

Apparatus

retort stand, rod and clamp,
  or tripod   items 503–505 or 511
1 small steel ball ($\frac{1}{2}$ cm diameter)
  or a marble   (12B)
1 large glass funnel (at least 20 cm diameter)

Procedure

Let the ball fall into a large glass funnel, which is firmly held vertically. Unfortunately, friction will affect the orbit and make it precess.

Select a ball which will fall right through the funnel.

Demonstration 27b
Illustration of elliptical orbit (*OPTIONAL*)

A small steel ball rolling on a distorted rubber sheet is a traditional method of showing various orbits according to the distortion imposed. Here too, friction retards the ball and makes a simple orbit precess.

Apparatus

steel balls, 3 mm diameter   item 131G
rubber sheeting from dental suppliers
toy hoop or embroidery frame
stands, rods, clamps, boss head   items 503–506

Preparation

Stretch rubber sheeting over a rigid horizontal circular frame and secure it tightly with tape. It should be stretched a little, equally in all directions.

Arrange a vertical metal rod over the sheet to be pushed down to depress the sheet into a curved well thus imitating, roughly, an inverse-square-force potential.

Procedure

Project a small steel ball across the sheet. Pupils watch its path. By choosing suitable initial conditions the ball can be made to describe an oval like an ellipse with one focus on the axis of the well.

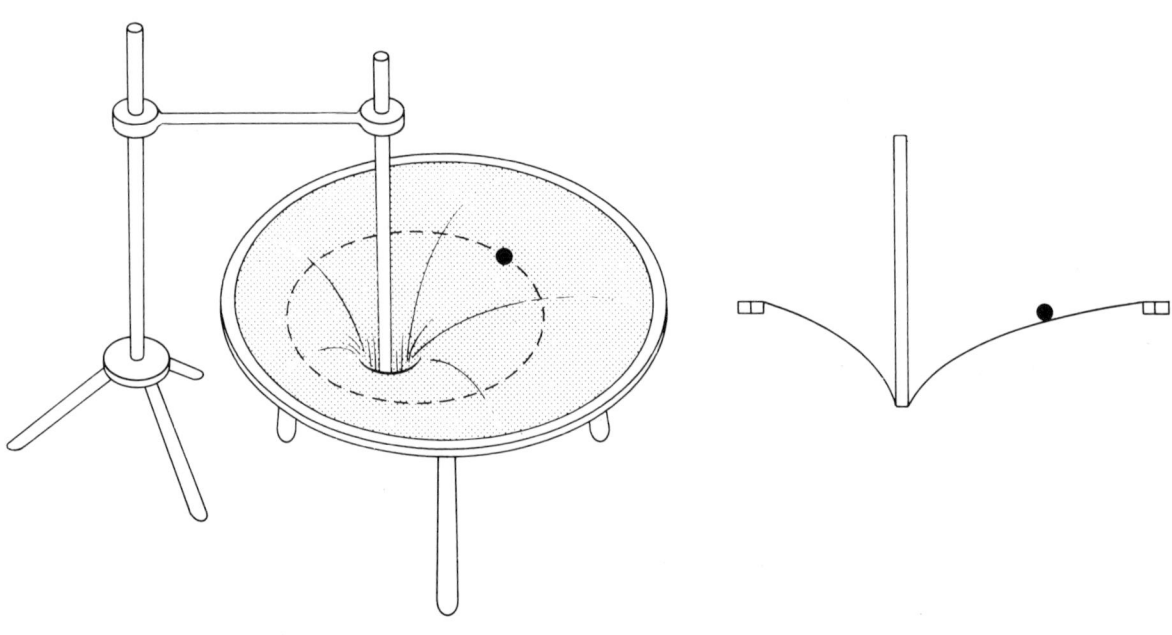

Class Experiment 28
Pulling a 'satellite' in

This gives a rough illustration of Kepler's Law II
– though the force is not strictly central.

Procedure

Pupils repeat **Experiment 2** at a gentle speed, then
allow the string to wrap round a finger.

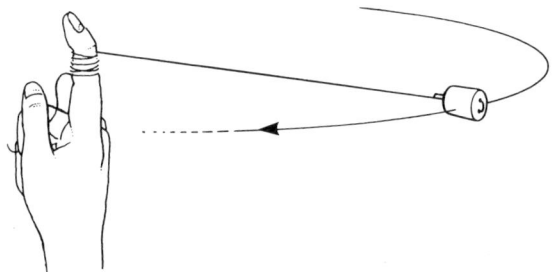

It is better to whirl the bung in a *vertical* circle.
Then the change of speed will seem more obvious,
though the motion is not quite so simple.

If the satellite is revolving slowly at first, the
change of speed is more apparent.

CHAPTER 5
The grand theory:
Universal gravitation

NOTES

Newton As they read, pupils may need a comment on the growth of Newton's knowledge stressing that his discoveries did not spring into his mind fully formed. He was interested in collecting knowledge and trying out ideas. And, at first, his views on motion did not take the final form of his laws, but were prejudiced by tradition.

The apple Pupils have probably heard the tale of the falling apple that suggested to Newton an extension of gravity. That tale seems to be both true and untrue. A friend recorded a conversation with Newton quite late in his life, in which Newton did tell the story of the apple. Yet, in his youth, about which Newton was reminiscing, he was concerned with other matters and such an early start on gravitation seems unlikely to some critics.

EXPERIMENTS

A useful demonstration of a planetary orbit can be improvised with the Edinburgh pucks kit.

Demonstration 29
Illustration of planetary motion using a dry ice puck (*OPTIONAL*)

Apparatus

Several arrangements have been proposed; one simple one is described here.

1 glass plate from Edinburgh pucks kit	item 95A
4 wedges	95B
2 brass ring pucks	95D
1 CO$_2$ cylinder with dry ice attachment	19/1 & 2
1 glass tube with fire-polished ends	
nylon monofilament thread	
1 bush of PTFE	
1 gantry (or two large stands and clamps)	161

The 'planet' is made up from two pucks, one on top of the other and held together with tape. This

increases the mass of the unit, which is desirable if unwanted friction is not to affect the motion.

The thread which tethers the puck in orbit should be nylon monofilament (from fishing equipment shops) to lessen friction at the centre.

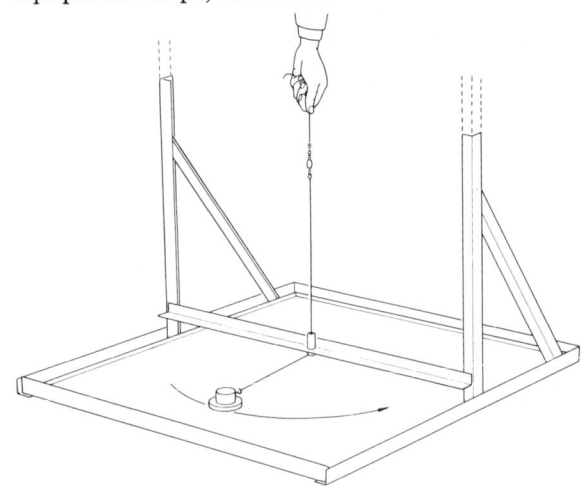

Preparation

As in all experiments with this system, the glass plate must be cleaned with methylated spirit or window cleaning fluid before use. And the planet should be kept in motion to avoid making cold patches on the glass.

The horizontal thread tethering the 'planet' must turn vertically at the centre of the glass sheet and be pulled by hand or by a demonstration spring balance. A short glass tube with fire-polished ends will serve for this. But a wider tube with a PTFE bushing does much better.

The gantry used for a camera can hold this tube, if an extra cross bar is installed at the bottom. Alternatively, two stands fitted with bosses and a long metal rod will serve.

Each revolution twists the thread. If this proves troublesome, interpose a small swivel like that used with fishing lines.

Procedure

When the 'planet' is floating on its cushion of dry

ice, start it moving in a circular orbit on the glass plate. Show the effect of increasing the speed. Change the radius of orbit while the 'planet' is in motion. It may be possible to show an oval orbit briefly, or at least to illustrate Kepler's Law II.

An oval orbit will 'precess' as a result of friction or any other unsuitable force law.

Class Experiment 30
Illustration of planetary motion with the centripetal force kit

Pupils repeat **Experiment 6** (Chapter 1), replacing the load with a pulling hand.

Apparatus

20 glass tubes, with rubber tubing	item 172A
20 rubber bungs (to act as 'planets')	172B
20 indicator cards	172E
ball of cord	172F

Pupils work in pairs and follow these instructions:

*   *   *   *   *

Take the rubber bung, thread, and glass tube from the centripetal force kit. Hold the free end of the thread in one hand and the tube in the other and whirl the bung round above your head. What

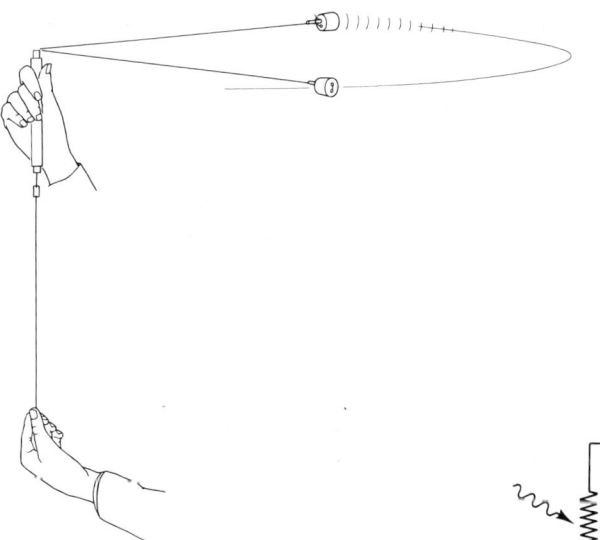

happens as you pull the thread a little harder? A little less hard?

*   *   *   *   *

Demonstration
Inverse Square Law with light
(An extra experiment to help the reading)

This is suggested as a quick demonstration to help pupils to realize the nature of an inverse square law and simple devices are now available. (Measurements with a traditional photometer would defeat this aim and might even be confusing.)

Apparatus

1 12-V 36-W lamp	item 73
1 lampholder (SBC) on base	74
1 transformer	27
1 photo-sensitive device†	
1 milliammeter, suitable for the device chosen	
small battery suitable for the device chosen (see below)	

† Light meters for photography do not read illumination directly; but the devices listed below will give direct readings.
(a) *Photo-transistor*, BPX 25, or a much less expensive one from component suppliers. This is already in use with the scaler as a timer. A demonstration needs a milliammeter with ranges 0–2·5 mA and 0–100 mA, both of them available in the demonstration equipment listed for our teaching. The currents are not exactly proportional to illumination yet a graph of I against $1/d^2$ is close to a straight line through the origin.
(b) *Light-dependent resistor*, ORP 12, from component suppliers. This is the least expensive and if pupils wish to try the experiment themselves with a small meter, 0–10 mA, it will give an excellent graph of I against $1/d^2$.
(c) *Solar cell*. Although the most expensive of the three, it is a welcome addition to the energy-conversion circus. It uses 0–10 mA meter, needs no battery, and gives a good graph.

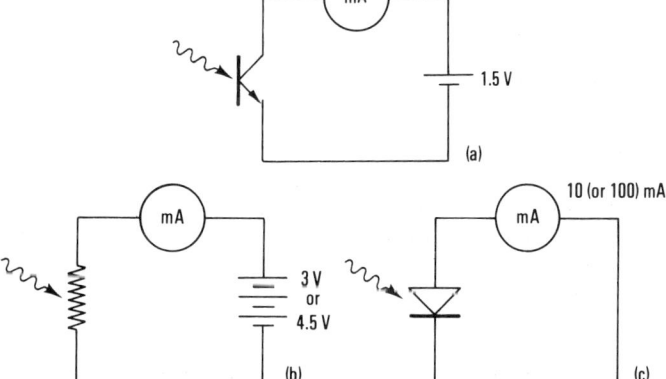

Procedure

Switch on the lamp in a darkened room. Place the photo-sensitive device 10 cm from the filament with its sensitive surface facing the lamp. Pupils read the milliammeter, which they may assume measures the intensity of light.

Then move the lamp until it is 15 cm from the device. Then 20 cm, 40 cm (and perhaps 50 cm).

Discuss the interpretation of the measurements. Plot the current, I (representing illumination), against $1/d^2$, where d is the measured distance from filament to sensitive surface.

Point out the general nature of an inverse square law. It must apply to any measured quantity which spreads out from a point in straight lines without any loss, such as light in clear air, electric field from a compact charge, sound from a small loud speaker in open space, gamma rays from a small source where there is no dense material, . . . but it does not apply to light in muddy water or light from a long fluorescent tube or traffic noise from a crowded road. (In the last two cases small individual sources radiate according to an inverse square law and it is only the geometry of extended sources that makes the law seem to fail.)

Experiment 31a
Rotation on a stool

Apparatus

1 rotating stool or platform   item 187

(A freely spinning stool or office chair may be used for these demonstrations. They should be oiled before use. A robust rotating platform can be made up from a motor-cycle wheel which is covered with a blockboard disk. A bicycle wheel is not strong enough.)

Procedure

Let as many pupils as wish try this in turn – personal experience is valuable. Pupils should follow these instructions:

$$\star \quad \star \quad \star \quad \star \quad \star$$

Sit on the stool holding a large book or weight in each hand.

Extend your arms sideways, and let a neighbour start you and the stool spinning.

Then draw in your arms and hold the two masses tightly to your chest.

Later, stretch out your arms again.

$$\star \quad \star \quad \star \quad \star \quad \star$$

Experiment 31b
Pupil spinning on his heel

Procedure

This is a simpler version of **Experiment 31a**, which a pupil can try for himself. He tries to spin on one heel for a short time.

If, in that time, he remains upright and holds a heavy book in his outstretched hands, he can feel the effect of moving the book in close to his body and out again.

With practice, it is possible to make two or three revolutions on one heel without the other foot coming to the rescue.

Demonstration 31X
Spinning demonstration with V-channel
(*OPTIONAL*)

Apparatus

2 steel balls or large marbles
   (diameter 2 cm or more)   item 96C or 131A, 12C
1 CO_2 ring puck   95C, D
1 glass sheet   95A
CO_2 cylinder and dry ice
   attachment   19/1, 19/2
V of angle channel†

† Construct the V with two pieces of angle channel arranged as in the sketch. Mount the V on a light wooden base.

Procedure

Level the glass plate and clean it. Place the mounted V on a CO_2 puck on the glass plate to provide a frictionless bearing for spinning.

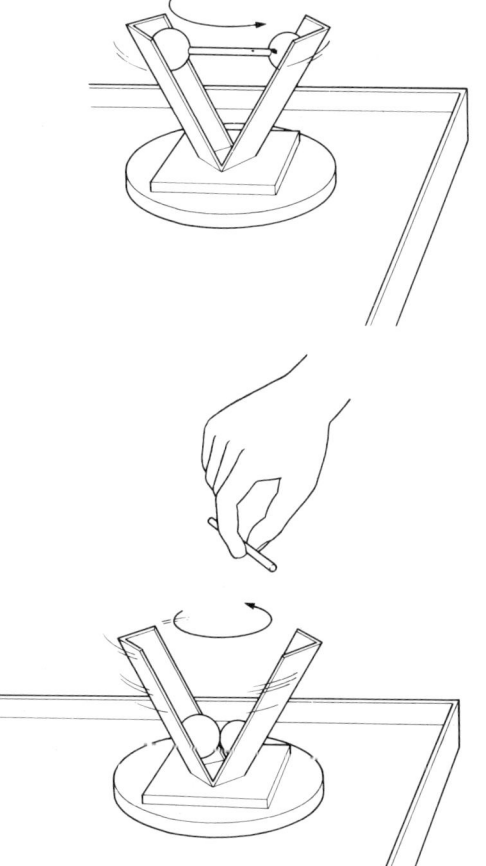

Load the V with two steel balls, one on each arm to represent two planets. Hold the balls high up by a piece of light metal tubing placed horizontally between them as a spacer.

Set the device spinning and ask pupils to note its speed. Then snatch the metal tube away. When the balls run down to the bottom of the V much closer to the axis, the system spins much faster.

Experiment 32a
Make your own model of a spinning Earth

Pupils follow these instructions:

    ★   ★   ★   ★   ★

Take a strip of very thin paper about 20 cm long and 2 cm wide. Make it into a loop by joining the two ends together with tape or paste. Very carefully, push the point of a pencil through the centre of the join – a small cut in the paper may help. Now push the pencil through this tightly fitting hole and across the loop. Where the point of the pencil meets the other side of the loop, make a hole just a little wider than the pencil. Push the pencil through that loosely fitting hole and then try rolling the pencil quickly to and fro between the palms of your hands.

    ★   ★   ★   ★   ★

Demonstration 32X
Model of the oblate Earth

Apparatus

1 large hollow rubber ball (from a toyshop)
1 metal rod (diameter about 6 mm)

Preparation

Drill small holes through the rubber at each end of a diameter. Slide a metal rod through the ball. The ball should fit loosely at one end, tightly at the other – so the diameter of one hole should be about half that of the rod, while the other is slightly larger than the rod.

Fit the end of the rod in the chuck of a hand-drill. Provide a stop at the other end of the rod – a small rubber bung will serve.

Support the drill with its axis of rotation vertical.

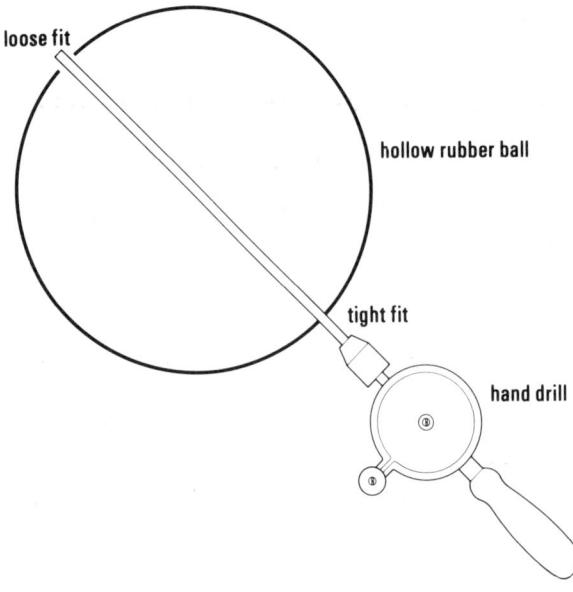

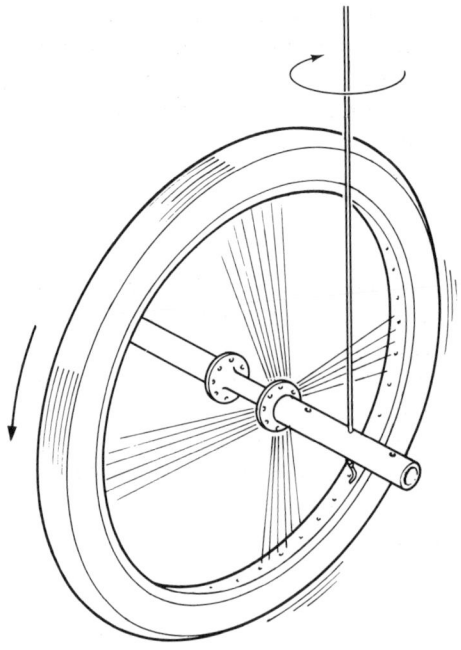

A 3-mm hole drilled through a diameter of one of the handles about 1 cm from the end takes a knotted cord that will suspend the wheel.

Procedure

Spin the ball fast.

Demonstration 33
Precession of a gyroscope

Apparatus

1 large gyroscope   item 186/1
$\frac{1}{2}$ m strong cord

Preparation

A bicycle wheel (46 cm diameter × 3 cm) does well – and even better if loaded with a tyre of lead strip.

Two 10-cm lengths of alkathene rod (or similar material) can be used for handles. The two rods are drilled out axially to a depth of about $2\frac{1}{2}$ cm with a 6-mm drill. They are then screwed firmly on the axle of the wheel to form handles.

Procedure

Hold the wheel vertically in one hand and set it rotating by a series of tangential blows with the palm of the other hand. When the wheel is rotating rapidly, it can be supported by the string, remaining vertical and precessing.

The wheel may also be set rotating by holding the rim against a motor-driven wheel.

Demonstration 34
Model to illustrate the precession of the Earth

Apparatus

1 large gyroscope   item 186/1
1 frame for the gyroscope   186/2
1 large rubber band
  nylon monofilament thread (from fishermen's shops)

Nylon monofilament is now available in many strengths. It does not make things spin when used for suspension, as ordinary thread and string do. Here any torque from the suspension would be unwelcome.

Procedure

Hang the frame which carries the rotating flywheel on a long thread, preferably from the ceiling. Spin the flywheel rapidly.

Arrange the spinning wheel with its axis tilted at about 45° from the vertical and across the pupils' line of sight. It will continue to spin with its axis keeping constant orientation thus. Then slip an elastic band over the hooks on the frame. Precession will start.

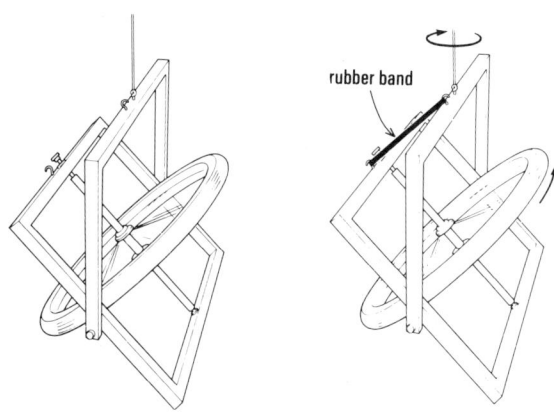

The pull of the rubber band represents the Sun's gravitational pull on the Earth's equatorial bulge. As soon as the rubber band is unhooked, precession stops.

ADDITIONAL NOTE

References for teachers

Teachers who wish to extend their own background knowledge of the field might find the following helpful:

Bronowski, J. *The ascent of man.* London, B.B.C. 1973.

Koestler, A. *The sleepwalkers.* London, Hutchinson, 1959 and Penguin Books, 1964.

Lewis, C. S. *The discarded universe.* Cambridge, C.U.P., 1964.

Mitton, J. and S. *Discovering astronomy.* London, Longman, 1979.

Rogers, E. M. *Physics for the inquiring mind.* London, O.U.P., 1960.

Ronan, C. A. *Discovering the universe.* London, Heinemann, 1972.

Toulmin S. and Goodfield J. *The fabric of the heavens.* London, Hutchinson, 1961.

OSCILLATIONS AND WAVES

We suggest that these four chapters should be treated quickly, as a change from astronomy and as preparation for atomic studies.

Chapter 6   Oscillations: simple harmonic motion

This is intended to give pupils qualitative acquaintance. The mathematical treatment, and the formal definition that would precede it, are outside the scope of the programme.

Chapter 7   Alternating currents

This continues the study of a.c. that started in Year 3. A generator of very low frequency a.c. provides special help in class experiments and some demonstrations. The d.c. power line (of Year 4) is now changed to a class experiment with low voltage alternating supply, followed by a demonstration with step-up and step-down transformers to show high voltage a.c. transmission.

Chapter 8   Waves

This takes up the discussion of diffraction and interference of water ripples and light waves in Year 3 and asks pupils to make an estimate of the wavelength of light with Young's fringes and with a diffraction grating.

Chapter 9   Spectra

Pupils see various spectra and are given a qualitative map of the electromagnetic spectrum.

CHAPTER 6

Oscillations:
Simple harmonic motion

LOOKING AT SIMPLE HARMONIC MOTION (S.H.M.)

A preliminary glance Let pupils look at three examples of simple harmonic motion, to see what they can learn without any previous lesson on the subject.

This equipment should enable each pupil to try each experiment by himself or herself. For this introduction to S.H.M. these experiments should not become demonstrations.

Class Experiment 35
Oscillations (a qualitative introduction)

Aim

To let pupils discover something about S.H.M. by their own unaided experimenting.

Apparatus

a.

6 retort stands, bosses, and clamps	item 503–506
6 G-clamps	44/2
6 lengths of thread (1 m)	
6 pairs of 5-cm metal (or wood) strips as jaws	121
6 1-kg masses to use as pendulum bobs	32

b.

6 expendable springs	2A
6 S-hooks	35
6 400-g hangers with masses	31/2
6 retort stands, bosses, and clamps	503–506
6 G-clamps	44/2
6 pieces of string, 25 cm	

c.

6 dynamics trolleys	106/1
12 expendable springs	2A
12 G-clamps	44/2
12 retort stands	503–504

Procedure

Explain to pupils that this is a chance to look at a special, important, type of motion. Leave them to find out anything they can about S.H.M. without telling them its name or what properties to expect.

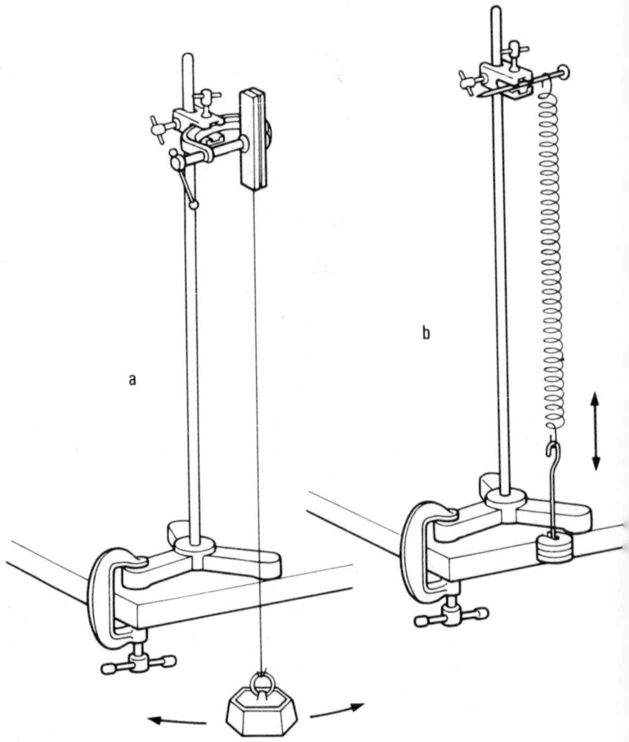

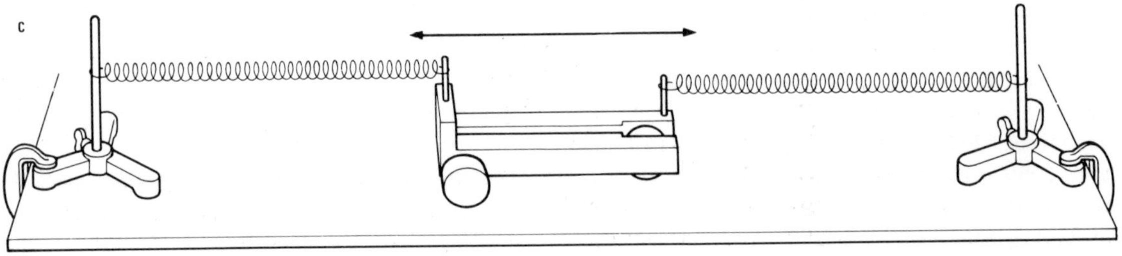

a *Pendulum* Pupils clamp the thread, with a kilogram mass hung on it, between a pair of 5-cm metal (or wood) strips which act as jaws. The jaws are held in a clamp attached to a retort stand fixed rigidly to the bench.

b. *Spring* Pupils suspend the spring from a retort stand and hang a load of 400 grams on it. They watch its up-and-down motion and try other loads.

Lateral motion (swinging as a pendulum with varying stretch) is interesting – and puzzling when its energy transfers to vertical motion – but it is too complicated to serve as a good example. Ask pupils to concentrate on the pure *vertical* oscillations. If they find those are soon obscured by transfer to lateral swings, suggest inserting a length of string (25 cm) between the bottom of the spring and the load. That will decrease the coupling between the two motions.

c. *Trolley* Clamp two small retort stands to the bench about 60 cm apart. Place the trolley between them and connect it to the stands with expendable springs. Pupils pull the trolley along and let go. The trolley can have extra loads.

Pupils listen as well as watch.

Conference Give, say, one class period (certainly not less) for pupils to find out what they can. Then at the beginning of the next period, ask what each found out.

Some will have discovered things that we do not normally list as essential characteristics of S.H.M. (examples: the dying-down of amplitude; the loaded spring's exchange of motion between vertical oscillations and pendulum swings). We should not condemn or even disregard these, but should accept them and even encourage further investigation. For example: is the dying-down of amplitude exponential? It would be easy enough to find out: and the question itself can be reworded in a much simpler form for a young experimenter. Or the observation could provoke the question: 'How could you make that worse?', leading to suggestions of paper sails, immersion in water, or perhaps to a discussion of mass.

Outcomes We have been insisting in the astronomy chapters that constancy is of the essence in scientific descriptions and laws. So it should not seem strange if a teacher now asks, 'Did you find something constant, in that motion?', or 'Did you

notice some behaviour which is the same in all three examples?' Yet for many pupils these questions would be too early; they might spoil the field now for future delights.

However some pupils will have found the isochronous property, or at least a strong hint of it. For the sake of others, they might be asked to keep it secret for a short while. And some pupils may have discovered some of the mass relationships – again, they might be asked to wait.

SIMPLE HARMONIC MOTION (S.H.M.)

S.H.M.: Important Explain that this is a very important type of motion. It is very common, and very useful. And it is the motion in musical instruments when they make a pure musical note. So we call it simple *harmonic* motion.

Explain that it is also important in more advanced physics because later, in A Level, we can calculate its frequency; and because, still later, we can analyse *any* repeating motion – tides, sound waves, motion of Moon, electron-waves in atoms, and many more – into a whole set of S.H.M.s. (If we ask for an earlier example, the pupil who says, 'Yes, Eudoxus,' should be awarded a prize!

{**Note** A load hung on a spring has two quite different types of possible motion: bouncing up and down and swinging sideways like a simple pendulum. The spring acts as a coupling agent, connecting these two motions. As in other 'coupled systems', energy is carried from one motion into the other, and then back to the first. The closer the two frequencies are to each other, the easier it is for the load, when moving with one motion, to excite the other motion.}

{If we work out the period of vertical bouncing of a load on a spring, we find it is equal to the period of a simple pendulum whose length is the *stretch* (the extension of the spring when that load is applied to the unloaded spring). Therefore, if the spring has a very small unloaded length, and most of the loaded length is stretch, the period of vertical motion is only a little shorter than the period of pendulum motion for the loaded spring. So the vertical motion soon transfers its energy to pendulum motion; then back again to vertical motion, etc.}

{If an experimenter tries to measure the period of the vertical motion, this is a very irritating phenomenon. The cure: insert a considerable length of string between the lower end of the spring and the load. This does not change the forces involved in vertical bouncing, so the period of bouncing is the same as before. But the period of pendulum-motion is now much greater, and transfer to that motion will happen much more slowly.}

Then give many demonstrations of S.H.M. Here are examples, some of which are sketched in *Pupils' Text*.

Demonstration 36
Examples of simple harmonic motion
Aim To give pupils a first look at many examples of S.H.M.

Suggested demonstrations

a. *Simple pendulum* Set three pendulums swinging – a simple pendulum side by side with one twice as long and one four times as long – without comment.

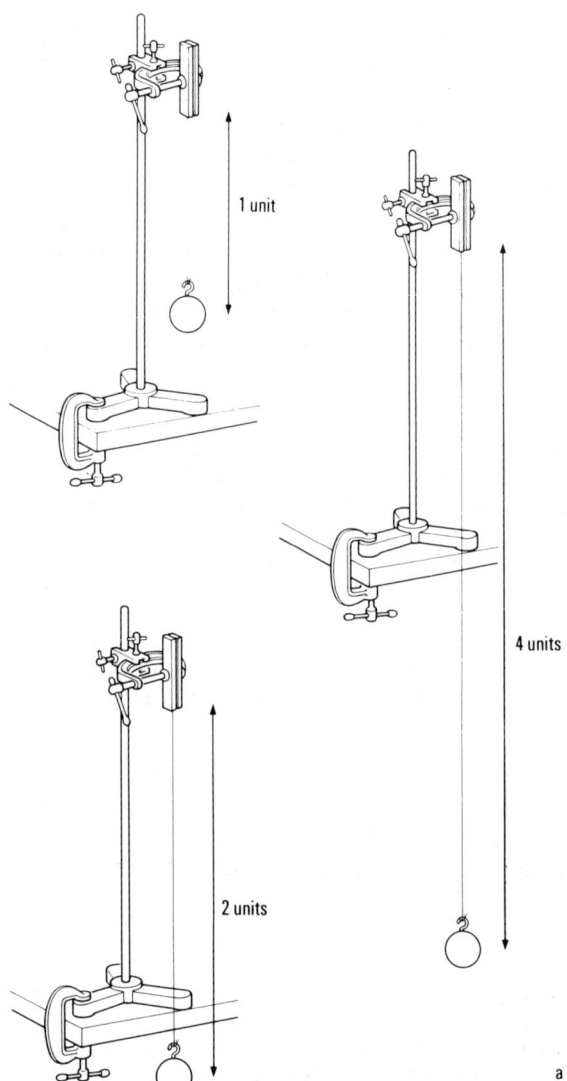

b. *Loaded beam* Clamp one end of a metre rule to the bench with blocks of wood or metal. Fix a boss or a G-clamp on the other end as a load. Let the metre rule perform vertical oscillations. Perhaps add a second load to show the effect of increasing mass.

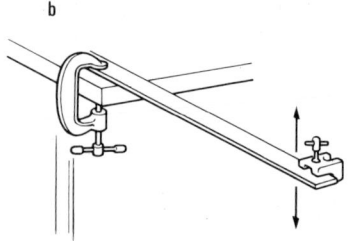

c. *Liquid in U-tube* Fill a U-tube half way up with water. (The tall mounted U-tube is unnecessarily big. One about 60 cm high will suffice.) Start the water oscillating by blowing into the tube.

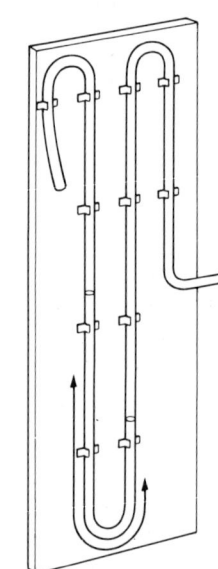

The motion is damped by fluid friction but is clearly isochronous. However, this is not the time to tell pupils that – their own discovery will be worth far more than clear instruction.

With a fast group, show two equal U-tubes, one half full of water, the other half full of mercury. (The mercury can safely be set in oscillation by a puff from a rubber bulb.) Let pupils compare the oscillations. Discuss the surprising result in terms of MASS and WEIGHT.

d. *Ball rolling in a bowl* Let a steel ball roll back and forth in a shallow spherical bowl, e.g., a large watch glass. ('Listen to the sound: what can you say about the motion?')

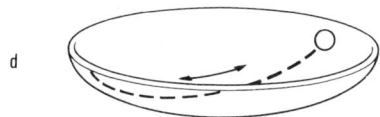

d

e. *Wig-wag* Clamp one end of the wig-wag to the bench. Load the other end and let it oscillate. Perhaps repeat with an additional load.

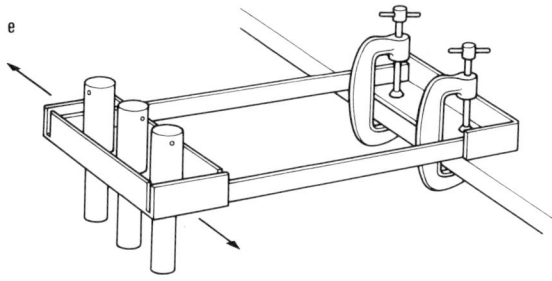

e

Demonstration 36f
An interesting comparison

Let a ball roll to and fro on three different shapes of rail. Pupils listen to the oscillations and try to decide which is simple harmonic.

Preparation

Bend three 50-cm lengths of plain metal strip (e.g., brass) 3 or 4 cm × 1 or 2 mm into three shapes: (I) a semi-circle; (II) a parabola; (III) a very shallow vee, with a short curve at the bottom. Attach them to a wooden back board (60 cm × 100 cm) by screws or small L-brackets underneath.

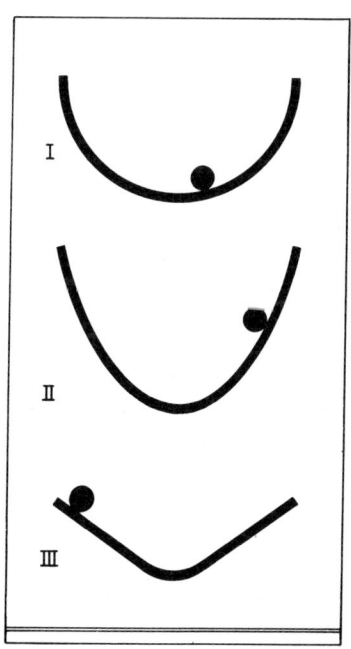

I

II

III

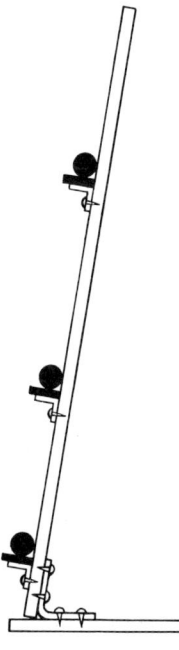

Set up the board on a wooden base almost vertical but leaning back a little (10° to 15°). Then a ball placed on one of the strips can roll to and fro without falling off. The noise it makes as it scrapes the backboard lets pupils judge the motion by ear – which is easier than by eye, because the sight of the decreasing amplitude is confusing.

Procedure

Let a large steel ball roll to and fro in each 'valley' in turn. Pupils listen to the motion. Avoid giving any hint of what they should expect, but ask them what knowledge they can extract *just from what they hear.* Leave them to draw conclusions.

In fact (though pupils should not be told this):
(I) has what sounds like isochronous motion;
(II) has a frequency that changes while the amplitude dies down;
(III) is not isochronous at all.

In (II) pupils may notice that the frequency changes, but few will notice that the motion becomes almost isochronous at small amplitudes. Offer a second chance, without prompting, and more will discover it. (To a mathematical inquirer this gives a hint of a 'circle of curvature' – the approximation for the nose of the parabola which, according to (I), will give S.H.M.)

In (III) the bend at the bottom may be a circular arc, but that is hardly a part of the general shape for (III). So it is best to take the ball away before it reaches small oscillations.

Time-graph of S.H.M. Show the graph of displacement in S.H.M. being plotted against time, and give the graph a name, 'sine curve'.

Demonstration 37
Time-graph of pendulum's simple harmonic motion

Apparatus

1 broomstick pendulum or equivalent   item 10F
1 roll of paper (width: 30 cm or more)
1 paint brush
ink

Procedure

Set up the massive broomstick pendulum used in previous years (or a loaded rod, or any other long massive pendulum). Attach a paint brush dipped in ink to the bob.

Set the pendulum swinging and pull a paper strip steadily across the floor under the pendulum, so that the brush makes a sinusoidal trace on it. (Pull the paper in a direction at right angles to the plane of oscillation.)

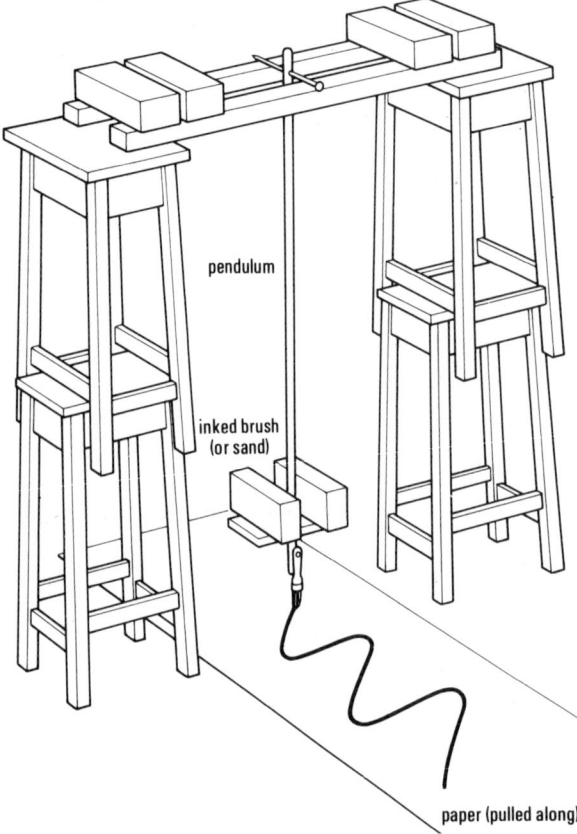

pendulum

inked brush
(or sand)

paper (pulled along)

If the paper is pulled faster, the sine wave is spread out more, in the same way as the trace on an oscilloscope when the time-base is speeded up.

Alternative marker Instead of a paint brush, hang a plastic cup on the bob. Make a hole (diameter 2 or 3 mm) in the bottom of the cup. Fill the cup with fine dry sand. The sand flows out and falls on the paper, and marks a line-graph of the bob's motion as the paper is pulled across the floor.

Musical S.H.M. Before discussing details of the motion or attempting a definition, let pupils see

examples of musical S.H.M., now that they can recognize it by its time-graph.

A tuning fork is necessarily massive and the amplitude of its motion is necessarily small when it is properly excited, so the classical textbook arrangement of a mirror stuck on one arm would make an unimpressive display. However, a strip of thin mica can be arranged to magnify the motion without much distortion and make a wonderful display.

Help from oscilloscopes Then pupils should see the wave-form of sound from a tuning fork shown by a microphone driving an oscilloscope.

We made a plea in Year 3, and a stronger plea in Year 4, for schools to provide small class oscilloscopes. Now in Year 5 we trust they will be available in many schools for pupils to use and watch at close range – one for every four pupils is adequate.

After offering a tuning fork to the microphone and C.R.O., pupils should try their own voices and other musical instruments.

Then, to continue their experimental learning about S.H.M. they should connect a small, isolated sample of the mains a.c. to the C.R.O. They have probably done this in an earlier Year, but the pattern has a new meaning now – it shows that the mains voltage has S.H.M. – and it forms a link with experiments in Chapter 7.

Demonstration 38
The motion of a tuning fork (OPTIONAL)

This is a fine demonstration, rewarding if the teacher can spend the time setting it up. It should be attempted if the apparatus is available.

Apparatus

1 tuning fork (256 Hz or more)	item 520
1 rubber hammer (rubber bung on a dowel)	
1 fractional horsepower motor†	150
1 small piece of thin mirror (5 mm square or smaller) ‡	
1 compact light source	21
1 converging lens (+7D)	112
1 L.T. variable voltage supply	59
1 white screen	102
small strip of thin mica (10 mm × 5 mm)‡	
1 plane mirror, 5 cm square or larger (mounted for motor)†	

† The motor is not essential. If the mirror is mounted on a free bearing it can be spun by hand well enough. The traditional cube with four mirrors is unnecessary: it might even divert attention by being 'special'.

‡ The small mirror on the mica strip is driven by the fork with much the same motion but larger amplitude. It reflects a small beam of light with considerable vertical motion which can be spread out horizontally by the large spinning mirror.

Procedure

Attach the small mirror to the mica strip near one end, with a contact adhesive. Attach the other end of the mica to one arm of the tuning fork. Clamp the fork firmly in a vertical position. Arrange the lens to form an image of the lamp filament 2 metres or more away. Place the tuning fork $\frac{1}{2}$ to 1 metre beyond the lens so that the mirror attached to it catches some of the light on its way to the screen.

Mount the single plane mirror so that it can rotate about a vertical axis. Drive it by the fractional horsepower motor, or turn it rapidly by hand. Place this mirror near the fork, where it can

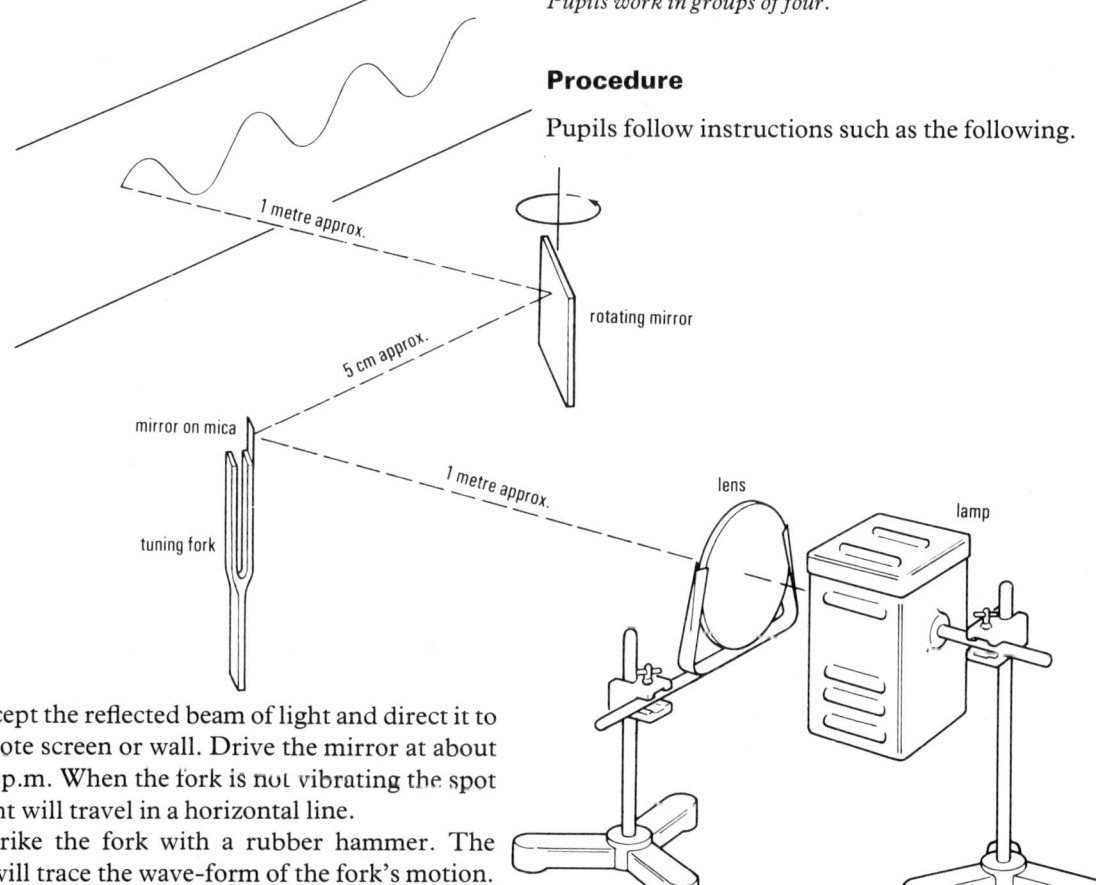

intercept the reflected beam of light and direct it to a remote screen or wall. Drive the mirror at about 600 r.p.m. When the fork is not vibrating the spot of light will travel in a horizontal line.

Strike the fork with a rubber hammer. The spot will trace the wave-form of the fork's motion.

Class Experiment 39X
Preliminary play with class oscilloscopes: learning the controls

The availability of class oscilloscopes depends on a school's planning and finances. They may have come into use in Year 3 or in Year 4 or they may arrive now in Year 5; but whenever pupils first use them it is important for all pupils, every member of each quartet, to learn to use the controls confidently. As in learning to ride a bicycle, personal experimenting is essential: lectures are no use and even printed sheets of instructions are of little help. Making acquaintance and gaining confidence will take at least a whole class period. So we insert this experiment here.

Apparatus

8 class oscilloscopes   item 158
8 low-voltage a.c. supplies

Pupils work in groups of four.

Procedure

Pupils follow instructions such as the following.

Join with a few partners to use a small oscilloscope. Take charge of the controls yourselves. Try the knobs and switches until you know how to use them.

Then connect a low-voltage a.c. supply to the Y-input of your oscilloscope.

Keep the time-base switched off at first; and then switch it on, set at range 2. The gain-setting should be about 1; the a.c./d.c. switch at d.c.

Try any changes you like. It is quite a robust instrument. The only important precaution is: *avoid leaving the spot at the same place for a long time, as it might damage the screen.*

Note Some teachers like to offer a small test to each pupil as soon as he or she feels ready for it. To give the test, turn all controls to zero or some to zero and some to maximum. Then 'Find the spot and make it draw a vertical line.' (For the next applicant who is ready for the same test, the question is different! 'Find the spot and make it draw a sine curve containing two wavelengths.' A new test for every applicant; and for any that the test finds unready, a different test after further experimenting.) There is a fine reward for success: a microphone to connect up and sing, talk, and whistle to.

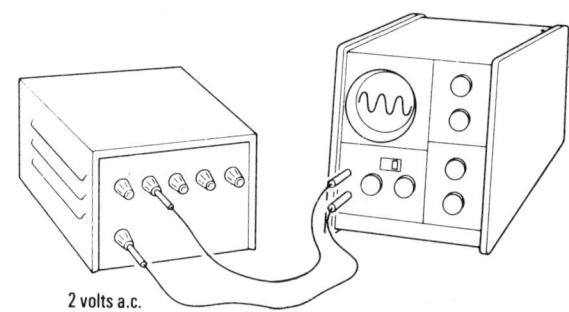

2 volts a.c.

at 1, the time base on range 2, and the a.c./d.c. switch to d.c. Ask them to adjust the variable control of the time base, and watch the pattern. What can be said about it?

Class Experiment 40
Wave forms of voices and other musical instruments shown electronically

Note If class oscilloscopes are not available – one for every four or five pupils – give a demonstration with the large oscilloscope (item 64).

Apparatus

8 class oscilloscopes	item 158
8 microphones (crystal)†	157
8 tuning forks	
8 small rubber hammers (rubber bung on a wooden dowel)	

† Carbon microphones from old telephones can be used instead. Each needs a 1.5 V cell and transformer in series with it. A small transformer designed to produce 6·3 V from the a.c. mains will do. Connect the output from the high-voltage windings to the input terminals of the oscilloscope.

Procedure

Connect the microphone to the input terminals of the oscilloscope. Set gain to maximum and the time base to the middle of range 2.

(*i*) *Tuning fork* Bring a vibrating tuning fork very close to the microphone, but do not let it touch. Adjust the time base to give a good display. If the fork fails to give a trace of sufficient amplitude, ask pupils to place the foot on a block of wood which will act as a sounding board. This will not give so pure a wave form.

Class Experiment 39
Wave form of mains a.c. voltage shown on class oscilloscope

Note If class oscilloscopes are not available – one for every four or five pupils – give a demonstration with the large oscilloscope (item 64), although the pupils have seen this before.

Apparatus

8 class oscilloscopes	item 158
8 low-voltage a.c. supplies	104 or 27

Procedure

Pupils should connect a supply of 1 or 2 V a.c. to the input terminals of the oscilloscope. Where the instrument used is the one described in Appendix 3 of *Teachers' Guide Year 4*, the gain should be set

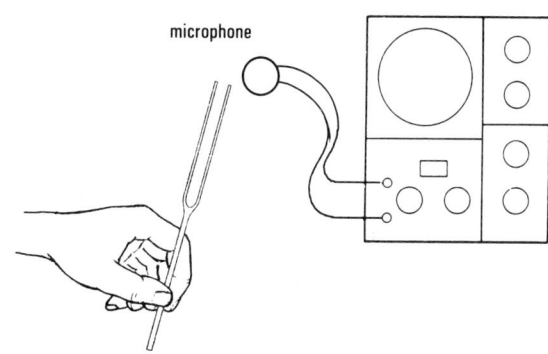

microphone

The microphone and oscilloscope show the wave form (or time-graph) of the sound waves in air that come from the fork. Is the motion in the waves S.H.M.?

Note: A tuning fork is designed to give a single pure musical note if it is struck carefully. Strike one arm near the top with a small rubber hammer, or bounce the fork gently against a rubber heel on a shoe. Avoid hitting it with hard metal – or even on the bench.

(ii) Voice Talk to the microphone. Sing gently to it. Try singing a vowel on one note: ah . . . , or ooh . . . etc. Are these *simple* harmonic motion?

(iii) Musical instruments If you play a portable musical instrument, bring it and see the wave form of its sound. Is that *simple* harmonic motion?

A THEORETICAL QUESTION

We have simply shown many examples of S.H.M., and pupils have seen the time-graphs of some.

Ask, 'Is the acceleration *constant*?'

Illuminate the matter by showing a large pendulum swinging.

Argue thus: 'Where is the bob moving *fastest*? Yes, at the centre. If it is moving *fastest* there, can it move any *faster* just beyond the centre? Can it be accelerating just there?'

Extending the discussion out to the extreme of amplitude when the bob is momentarily at rest there will promote a lot of argument. With a fast group this can reveal, and cure, some confusions between position, velocity, and acceleration, or doubts over maxima and rates of change.

With slower groups this argument is worrying and should be avoided.

(With a very long pendulum it is possible to make measurements of velocity with the scaler to count milliseconds using a photo-transistor, but this would soon become tedious for most pupils. We mention it only as an advanced optional project for a few keen pupils.)

DEFINING S.H.M.?

A formal definition does not seem necessary or fruitful here; but end the discussion above by pointing out the changes of acceleration from maximum inward at each end, to zero at the centre. We can tell pupils that when that is put in proper mathematical form we can calculate the period of any S.H.M. from its two characteristic factors:

   a mass factor (the load on the spring),
   a springiness (or force) factor (the slope of the spring's Hooke's Law graph).

Unfortunately, in the example that pupils consider the most common, a pendulum, that story of two factors seems to fail! It does not really fail, but since the springiness factor is a *fraction of mg* and the mass factor is m the mass disappears in the prediction of T.

A note on pendulum measurements

Pupils might be given pendulums to investigate. But we should not ask them to derive a measure of g unless they are so able that they can follow the full derivation of the formula. If we issued the formula to pupils who have little idea of its derivation, and expected them to make such measurements with its help, we might well give science a poor reputation. Those pupils who proceed to A-level physics may have an opportunity there to use the formula with full understanding. But we should remember that detailed precise measurements of g do not seem so important as they once did.

Work with pendulums in a teaching programme can take several forms.

a. *Practical work to demonstrate the relationships, or 'verify' the laws*

Although many pupils welcome routine work with definite instructions and a clear outcome, this form in which the answer is stated first will not give a good example of scientific work but rather show it as an obedient carrying-out of duties. In retrospect it will be dull.

b. *Training in techniques of timing and observing*

Training does not transfer easily; it does not spread to other fields of science, or to life in general, except when its potential value makes a strong impression. For most pupils, training with pendulums will be wasted.

c. *A scientific investigation*

Even though pupils know that the teacher is well aware of the 'answer' – that he has some 'formula' which tells him how pendulums behave – they can do genuine scientific work if they regard the experiment as an investigation. The experiment itself may be almost the same as 'verifying the formula' yet with a different attitude in the instructions. Put it to pupils that they are simply asked to find out something – not to show that what they have already been told is true. To set the stage, we do say what should be measured (e.g. that many swings should be timed) or what relationships should be investigated (e.g., *T vs L*); but we do *not* say what results are to be expected.

d. *An accurate measurement of* g

After an investigation, the formula could be given to keen pupils who would enjoy extracting a value for *g*. But such an indirect measurement of *g* is of little importance to-day.

Class Experiment 41
Look for rules for pendulums

Discussion

Give the period of the pendulum as the thing to be investigated and discuss briefly with pupils the physical conditions that might affect the period. Ask for suggestions. Pupils know that a longer pendulum takes more time to swing to and fro, so they are likely to suggest LENGTH as one factor affecting period.

Encourage pupils to make further suggestions and hope to hear MASS of the bob; AMPLITUDE (which we describe with a sketch as the MAXIMUM ANGLE OF SWING FROM THE VERTICAL and we postpone that at first). Also, perhaps, *g*.

Let pupils start by investigating the effect of mass as a factor. They make timings for a long pendulum with a small bob and then with a large bob. They can keep the length factor constant – as they should while they investigate something else – but unfortunately they cannot keep the amplitude constant. They may come back with the

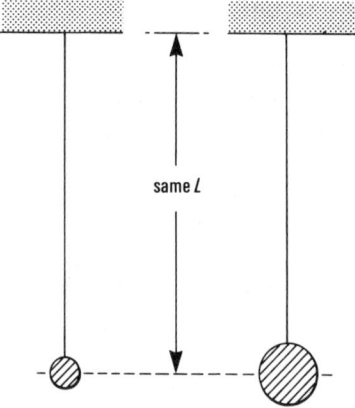

same *L*

complaint that the swings are dying down in amplitude, although they are not trying to investigate that change. If so, it is best to say, 'Don't worry; forget about that for now and continue for timing. You can investigate that dying-down effect later.'

{Of course, with mature students we should elicit the question of amplitude effect and point out that the *first* investigation should logically be of *period* versus *amplitude*. When that has been settled small amplitudes can safely be used for the other investigations although they die down visibly as we know, and as pupils will discover. However, for young beginners that logic is not a delight but a puzzling argument which spoils the beginning of the investigation.}

Procedure

a. *Period for different masses of bob* Pupils first try two different bobs on a long pendulum. The timing of these will show that period is independent of mass. Although that is not the simplest way of showing it, it is probably the most impressive. (The simplest: start two pendulums side by side and watch them swing in unison!)

When pupils come back surprised with the result that the period with a heavy bob is the same as the period with the light one, tease them, saying: 'You could have predicted that from something you already knew'. If they cannot guess, finally point out that this is a case of falling bodies of different masses. The string of the pendulum does not help the acceleration of the bob along its arc or hinder it, so we should expect this to be a diluted form of the leaning-tower demonstration.

b. *Period for various lengths* Making careful measurements for pendulums of different lengths,

78

and plotting graphs, can easily become a long tedious business – and, for pupils at this stage, a rather pointless one. Instead of that, we suggest that each pair of pupils should make *one* measurement, repeated once or twice to obtain a fairly reliable value of period for one measured length.

To provide a useful range of results, assign different lengths to different pupils beforehand.

Then plot a large communal graph on behalf of all pupil investigators. If the initials of the pupils who made the measurements are pencilled against each plotted point, some useful (and heated) discussions may ensue.

A straight-line graph? To a scientist, a survey of the measurements might well reveal the relation $T^2 \propto L$. A fast group might discover that if asked leading questions.

Ask what should be plotted if we hope for a straight-line graph. (See the discussion below of straight-line graphs, their advantage and interpretation.) Ask about plotting L on the x axis and $1/T$, $\sqrt{T}$, or T^2 on the y axis. Testing each of those by a full plot would take too much time. Instead, show pupils how to make a quick feasibility test by arithmetic. Choose from the measurements (or from the first graph) the values of T for any length and for double that length. Then, for a straight line through the origin, the appropriate function of *T must also double when L is doubled.*

Fictitious example:

L	T	$1/T$	$\sqrt{T}$	T^2
40 cm	1·27 s	0·79	1·13	1·61
81 cm	1·80 s	0·56	1·34	3·24

Therefore try plotting T^2 against L.

Then plot a large graph of T^2 versus L for all pupils' measurements. If the plotted points seem to lie close to a straight line, use a taut thread to look for the 'best straight line' and draw it. If pupils think the line passes near to the origin, they may prefer to force the chosen straight line to pass through the origin. With that choice, the straight line will offer an answer to the question, 'How close to the simple law $T^2 \propto L$ is the behaviour of the pendulums in our lab?'

c. *Period for different amplitudes* The *Pupils' Text* asks pupils to *guess* a third possible factor. So when it comes to (**c**) the *Text* does not mention amplitude but says 'Discuss with your teacher.'

If amplitude has not been suggested, make the suggestion and ask pupils to make measurements with a long pendulum swinging with amplitudes about 40°, 20°, 10°. The most skilful among the experimenters may like to try 60° or 5°.

The large amplitudes die down quickly and only rough measurements of a few swings will be possible. Explain to pupils working with these that they must be content with rough measurements and they should let the amplitude run from, say, 70° to 50° for their '60°' measurement.

Since the period is the same (within the accuracy of these measurements) for all small angles, and even at 60° it is only 7 per cent greater, it is hardly safe to combine the measurements from pupils with different pendulums: for example, pendulums meant to be equal but really of lengths 100 cm and 101 cm would differ enough in period to swamp the difference between 40° and 10° amplitudes. So each pair of pupils should try only three amplitudes, otherwise the experiment would be very long without being much more fruitful.

In retrospect pupils will have been safe in using amplitudes of 10° even if they died down to 5° during a measured batch. Point this out.

Also point out the value of pendulums in clocks.

{**Errors in pendulum measurements** If our pupils in a fast group understood the random walk argument in Year 4, we might mention it again here. We point out that if we make a measurement many times and take the average, we are really dealing with the result of a large number of errors like steps in a random walk. Therefore by taking ten times as many sets of measurements and averaging them, we do not reduce our likely error by a factor of 10 but only by the square root of 10.}

Notes on seeking a straight-line graph

We might remind pupils that the advantages of a straight-line graph are:

1. It is easily drawn with a ruler.

2. When we choose the 'best straight line', we take a 'weighted average' of our measurements, giving less weight to points that seem out of line with the rest. That is often good scientific procedure though it may be dangerous. (Wishful thinking? Rigging the ballot?)

3. *If we draw a straight line* through the origin, *it represents direct proportionality between the two things plotted.* And if our plotted points lie close to such a line, we can say that our measurements show that the behaviour of our experiment is close to that proportionality.

{With a fast group, one might expound the logic of this fully. Pupils should see that the straight line we draw expresses an *ideal*, perfect proportionality. Our points express the *facts* of our experimental observations. When we compare the points with the *line*, we are comparing the *facts* with a *simple proportionality law that we hope for.*}

{There was an important example in earlier years when pupils measured the motion of a trolley running down a hill, or a trolley pulled by a constant force. We hoped to find that this is a case of constant acceleration. If we plotted distance, s, against t^2, where t was the total time of travel from rest, we hoped to get a straight line. A straight line drawn through the origin on that graph has the equation $s = (\text{constant}) t^2$.}

{We may draw such a line whenever we *know* that s is directly proportional to t^2. In fact, we do know that s is proportional to t^2 for any case of *constant acceleration from rest*. We know that through simple, reliable mathematics, leading from the statement:

$$\frac{\text{CHANGE OF VELOCITY}}{\text{TIME TAKEN}} = \text{ACCELERATION (which}$$

we often find is constant) to the result $s = \frac{1}{2}at^2$ providing *a is* constant.}

{No experiment is necessary to show that.}

{*IF a* is constant, *THEN* $s = \frac{1}{2}at^2$, because logic does that. Therefore, no experiment is necessary to show that the graph of s versus t^2 is a straight line through the origin for constant acceleration from rest. Then, why do we plot the graph? What are we doing when we draw the straight line among our plotted points? *We are trying to find out whether our trolley really moved with constant acceleration.* We know the line is true for constant acceleration; so, by seeing how close our points come to the line, we are seeing how close our trolley's measured motion comes to constant acceleration.}

{In many experiments we can find or construct some functions of the measurements which, when plotted, will bring the plotted points near a straight line; but the 'best straight line' may fail to go through the origin. In that case, drawing a straight line, and looking to see how close the points are to it, is not asking whether we have a case of direct proportionality, $y = kx$. Yet we are still asking whether there is a simple linear relationship $y = kx + c$. To us as physicists that relationship is almost as simple and interesting as simple proportionality, but pupils either find it less clear and simple or think it is just the same as proportionality, and we need to teach the difference – if we find pupils ready to appreciate it easily.}

{In some experiments, all our measurements of one quantity are wrong by a constant amount. (For example, in a pendulum investigation of T versus L all the lengths may be too small because we used the length of thread for L and forgot to allow for the bob.) Then we should choose our functions for plotting so that the measurement with the constant defect is left untouched. (In our example, plotting T^2 against L will still give a straight line, if every value of L is too short by the bob's radius, but plotting T against $\sqrt{L}$ will *not* give a straight line.}

{Then, if our graph in the ideal case would be a straight line through the origin, the intersect where the best line fails to pass through the origin may give us valuable information. (In our example, it simply tells us the radius of the bob – rather inaccurately.)}

{One of the most far-reaching examples is the graph of PRESSURE of gas in a flask (constant volume) versus TEMPERATURE. The intersect on the temperature axis gives an estimate for absolute zero.}

CHAPTER 7

Alternating currents

SYNOPSIS OF PROGRAMME

{**Dynamos, meters, and oscilloscopes** Return to the electromagnetic kit for some 'catching-up' experiments. Resume the study of the simple dynamo and let pupils try for themselves the output of their own d.c. and a.c. dynamos on a moving-coil meter; then on an oscilloscope. Show the bicycle dynamo and try that on a centre-zero meter at low speeds; then on an oscilloscope.}

{**Mains a.c.** Then let them try a sample of 50 Hz a.c. derived from the mains on a class oscilloscope.}

{**Characteristics of a.c.** It would be easy to leave a.c. as a slightly mysterious version of the direct currents that pupils dealt with in earlier experiments. But a.c. is the household form of supply and is far more economical in distribution because of the efficiency and simplicity of transformers. So pupils are likely to be interested in the characteristics of a.c.: the obvious ones, such as giving the same heating effect as a direct current and failing to move a d.c. ammeter visibly; we may also show for interest some surprising characteristics involving phase-differences. We therefore should give some teaching of a.c. to all classes, expanding it for faster groups.}

{**Oscilloscope** Again show the wave form of an alternating voltage. Give some useful names: peak voltage, average voltage (which is zero!). Point out that there is a voltage almost all the time and ask for suggestions for a way to specify something useful instead of the plain average of zero. Explain how we use r.m.s. values.}

{With a fast group, suggest the idea of squaring the voltage V, finding the average value of V^2, and taking the square root of that. We might even remind pupils that such an average is what we really calculated in Year 4 for the speed of air molecules.}

{For a slower group, we might just say that we can take an average of the upper half of a cycle which is positive, and arrange to take a similar positive average for the lower half.}

{It is helpful to make an oscilloscope show the alternating wave form and then sketch an average value in its face with a felt-tipped pen.}

{**Ohm's Law** Let pupils see the behaviour of a resistor with a.c.}

{**Power or transmission line** Pupils return to the model transmission line of an earlier Year. Now they try it with low voltage a.c. They then see transformers used to transmit power at high voltage.}

{**Current and voltage for capacitor and inductor?** Teachers with a fast group might feel tempted to go further and show phase changes with an oscilloscope for voltages applied to a capacitor or an inductor. But that would prove a much harder matter to understand than one would expect. It is much wiser to make this study part of the work with 'slow a.c.'.}

{**Slow a.c.** Although mature physicists can easily imagine peak values and discuss the meaning of r.m.s. values, these things are much easier for beginners if they see them with very slow a.c. and watch the moving pointers of meters.}

{Demonstrations with slow a.c. have long been a tradition. We hope that everyone teaching alternating currents at this stage will not only have demonstrations of slow a.c. but will have enough equipment to put some of these experiments in the hands of pupils. Several separate groups of pupils could be fed by the same slow a.c. generator.}

EXPERIMENTS FOR CATCHING UP

These are suggested as reminders, to refresh memories of class experiments on dynamos in Year 3.

Pupils who missed that work entirely will need a more thorough start. We suggest that any such pupils should use *Pupils' Text Year 3* and go quickly through Year 3 experiments 99, 102, 103, 104, 105, 106, 108a, 108b. Several of these are referred to in this chapter for catching up.

†Class Experiment 42
Simple d.c. and a.c. generators

Apparatus

1 electromagnetic kit	item	92
16 galvanometers		180
8 class oscilloscopes		158

† Denotes revision of work of an earlier year.

Procedure

The diagram given in *Pupils' Text Year 5* may be adequate reminder for those who have built the motor before. Pupils should work in pairs and, if necessary, can be referred to the instructions in *Pupils' Text Year 3*.

a. Pupils should first construct the model electric motor and make it run. Suggest a competition for making the best motor. A small load hung on a thread from the axle tube will enable the owners to estimate the motor's brake power, and that can be used as a measure of success.

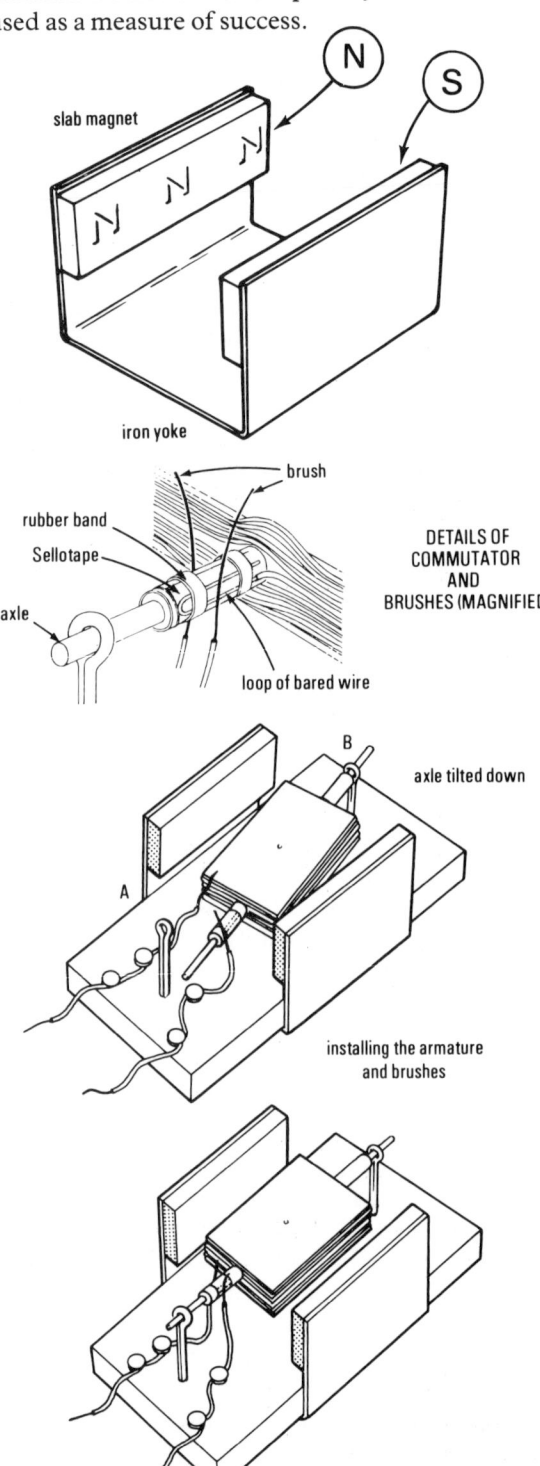

slab magnet

iron yoke

brush
rubber band
Sellotape
axle
loop of bared wire

DETAILS OF COMMUTATOR AND BRUSHES (MAGNIFIED)

parts for model dynamo

axle tube

Sellotape for insulation

10 turns of wire

axle tilted down

B

A

installing the armature and brushes

b. Then, treating their motor as a dynamo, pupils connect it first to a galvanometer, then to a class oscilloscope using maximum gain. With the CRO they should drive the model fast with a thread wrapped round the axle tube and 'sawed' to and fro.

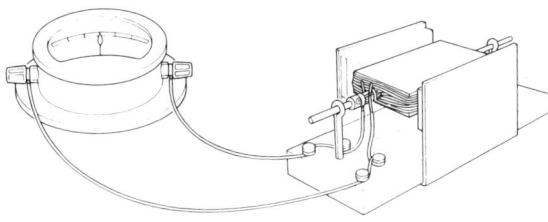

c. Pupils re-build their motor as an a.c. generator and then try its output on a class oscilloscope.

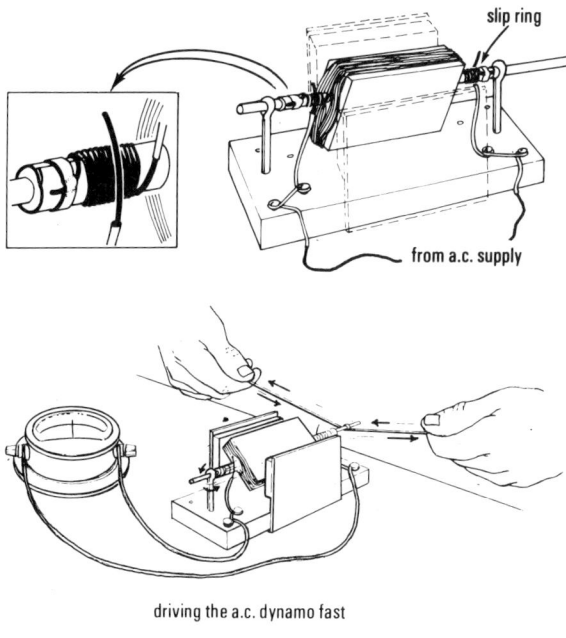

driving the a.c. dynamo fast

†**Demonstration 43**
Bicycle dynamo

Show this now only if pupils missed it in Year 3. Otherwise, just offer it as a pupils' demonstration for revision.

Apparatus

1 bicycle dynamo assembly	item 103
1 demonstration meter	
(or galvanometer)	70 (or 180)
1 d.c. dial 2.5–0–2.5 mA	71/4
1 demonstration oscilloscope	64
(or class oscilloscope)	(158)

Procedure

a. Connect the output from the generator to the meter and turn the handle slowly.

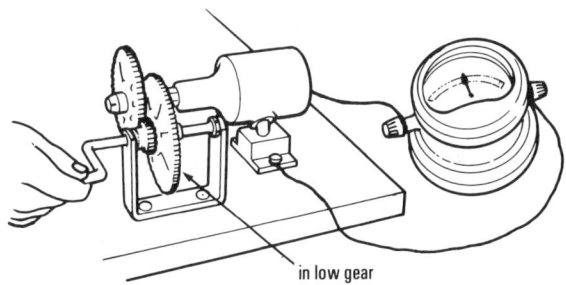

in low gear

Increase the speed. At high speeds the pointer merely vibrates over a small range about zero.

Drive the dynamo still faster with the other gear, and use it to light a 2·5-V lamp. (Remember to disconnect the meter first.)

b. Connect the output from the generator to the input of the oscilloscope, with maximum gain on its amplifier. Drive the dynamo slowly, with the time-base of the CRO switched off. The spot moves slowly up and down.

Then switch the time-base to slow speed and centre the trace with the X-shift, keeping the vertical gain at maximum. Gradually speed up the time-base.

Reduce the vertical gain to about 2 volts/cm and drive the dynamo fast. With the time-base at 10 ms/cm, the trace will show the wave form.

Note The wave form is not sinusoidal; the bicycle dynamo was designed for efficiency and not for teaching purposes. Other generators can be found which give a more nearly sinusoidal wave form, but there is greater value here in using a generator as familiar as the bicycle dynamo.

†Class Experiment 44
Dynamo effect: magnet and coil wound on iron core

Apparatus

From electromagnetic kit: item 92
   PVC-covered copper wire   92X
   16 C-cores   92G
   16 ticonal magnets   92A
   16 galvanometers   180

Pupils work in pairs. Each pair needs about 2 m of wire.

Procedure

Pupils follow these instructions.

*   *   *   *   *

Wind a coil of about 25 turns on one arm of the C-core and connect the coil to a galvanometer by long leads.

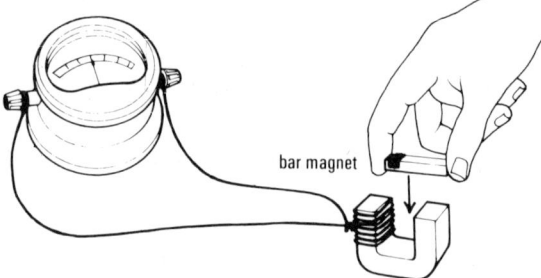

bar magnet

Place a magnet across the ends of the core, and watch the effect. Then remove the magnet, and watch the effect.

*   *   *   *   *

Induced voltages Using that experiment as an example, summarize pupils' experience of electromagnetic induction in Year 3. This Year's *Pupils' Text* discusses the effects and concludes: there is an induced voltage while a magnet is moving towards a coil or away from it, when a coil or any other wire moves across a magnetic field, or when the magnetic field in the region of a coil is growing stronger or weaker. It is *motions* or *changes* of a magnetic field relative to a wire that induce voltages in the wire. This was Faraday's discovery that led to the development of dynamos and power stations, when before there had been only batteries.

†Class Experiment 45
Simple transformer with d.c. supply and switch

Apparatus

From electromagnetic kit: item 92
   insulated copper wire
   16 pairs of C-cores and clips   92G
  16 galvanometers   180
  16 dry cells   52B

Note Low-voltage power units could be used in place of the dry cells, but the ripple on the d.c. output will probably lead to confusion. There will be a galvanometer deflection even when the electromagnet is left switched on. Dry cells are therefore much better.

Pupils work in pairs, each pair needing about 4 m of wire.

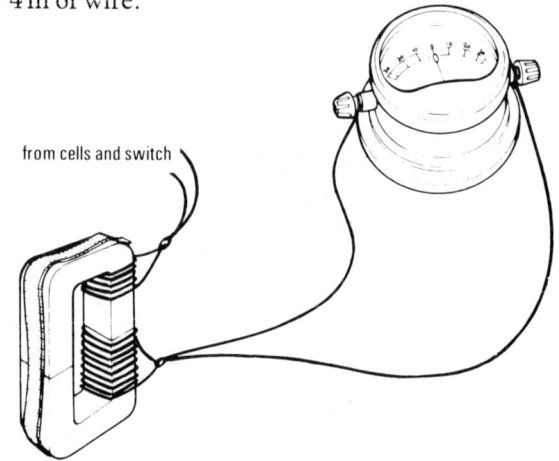

from cells and switch

Procedure

Pupils follow these instructions

*   *   *   *   *

Take the C-core used in **Experiment 44** to form the secondary of a transformer, keeping it connected to the galvanometer.

Wind 10 turns of wire round another C-core to form the primary.

Clip the two C-cores together.

Touch the two ends of the primary coil on to the terminals of a dry cell. Watch what happens. When does it happen?

Be careful not to leave the dry cell on too long.

★   ★   ★   ★   ★

Class Experiment 46
Simple transformer with a.c. supply

Apparatus

From electromagnetic kit:	item 92
16 pairs of C-cores with clips	92G
16 MES lamps (2·5 V, 0·3 A)	92R
16 MES holders	92T
16 low-voltage a.c. supplies (1 volt)	104

Procedure

Pupils follow these instructions.

★   ★   ★   ★   ★

Use the same arrangement of two cores and coils as in the previous experiment; but replace the dry cell by an a.c. supply.

Connect the ends of the secondary coil (25 turns) to a lampholder with a 2.5 V lamp in it.

Connect the primary (10 turns) to 1 V a.c. supply.

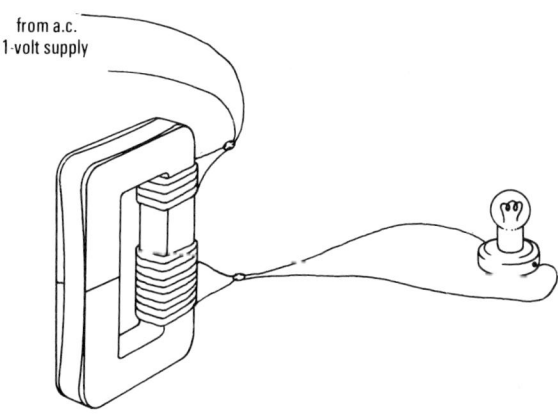

from a.c. 1-volt supply

If you like, see how well a lamp will light without the help of this 25:10 'step-up' transformer. You need not disconnect your lamp; just connect another, equal, lamp straight across the low voltage a.c. supply. Does that second lamp glow as brightly? To show that this is not due to some difference between the two lamps, interchange them.

★   ★   ★   ★   ★

Pupil's Text describes the large contribution made by the iron core.

Demonstration 47
Winding a transformer turn by turn

This was suggested as 'optional but desirable' in Year 3. Unless pupils saw it then, now is the right time for it as an essential demonstration.

Apparatus

1 demountable transformer	item 147
with 1 coil of 3600 turns	147G
1 MES lamp, 2·5 V	92R
1 MES holder	92T
spare lamp	92R

flexible, well insulated wire (4 m or more)†

The demountable transformer is necessary for this demonstration.

† For the flexible wire, one of the wires from twin lamp flex is better than the PVC insulated wire.

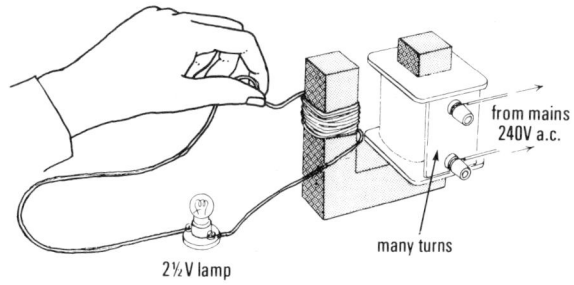

from mains 240V a.c.

many turns

2½ V lamp

Procedure

Place the 3600-turn coil on one leg of the laminated U-core of the demountable transformer. Connect it to the a.c. mains.

Connect the long flexible lead to the lampholder with the lamp.

Switch on the mains and wind the long wire *turn by turn* round the other leg of the U-core. As more and more turns are wound on, the lamp begins to glow and then gets brighter and brighter. At least 25 turns will be necessary for this.

Provided a spare lamp is available, try placing the I-yoke across the top of the U-core.

Pupils' Text discusses the role of numbers of turns in primary and secondary. It also explains why the core is laminated.

The demountable transformer and its coils

There have long been demountable or dissectable transformers for teaching, usually in an outfit for illustrating all electromagnetism by lecture experiments.

Although we place most emphasis on pupils' own work in class experiments, there is occasionally an important experiment which might be dangerous in pupils' hands, or one that needs apparatus that is too expensive for multiple sets. One of these is a simple transformer with interchangeable coils, large enough to be watched by a whole class, and designed to meet several needs in our Years 3, 4, and 5.

The core should be a laminated U, at least 11·5 cm wide by 12 cm high with a laminated I-yoke that is easily removed and restored. It should have a cross section at least 3 cm × 3 cm.

The arrangement for clamping the core to a base-board and for clamping the yoke to the core (seldom necessary) should be as simple and unobtrusive as possible. Supporting clamps should not bulk large in the appearance of what is meant to be a clear 'skeleton' device.

The coils, which should be wound professionally and given good terminals, should have numbers of turns chosen to fit our needs economically. Those needs have changed a little. Originally, we asked for coils of 300 turns, 600, 600, 1200, 3600, 12000 turns. Schools that already have those will find they meet all needs. Schools now planning to buy the transformer are advised to buy the following coils: 300 turns, 300, 600, 3600.

With two 300-turn coils (and only one 600) the 1200- and 12000-turn coils will not be needed. For an inductance with slow a.c., the 12000-turn coil is better replaced by the 'high inductance coil', 1100 turns, (designed for A Level) on a pair of C-cores.

Mains a.c. Now, if not before, pupils should try a sample from the mains on their oscilloscopes. This is an important part of the work in this chapter; so, unless they have already tried it in Chapter 6, they should certainly try it now.

Class Experiment or Demonstration 39 (repeated) Wave form of mains a.c. shown on an oscilloscope

See pages 75 to 76 for details of the class experiment suggested and for the 'preliminary play' with class oscilloscopes needed by pupils using them for the first time.

Any transformer from mains to an isolated secondary giving a few volts will serve here but a 'Variac' should not be used.

For a demonstration, a transformer that makes its function clear would be best. Either use the coils that fit the C-cores used in the a.c. power line demonstration (items 127, 128, and 92G) or, for a more impressive but equally clear transformer, the demountable transformer (item 147 – see the note above). Install coils on the core and complete the core with the yoke. Use the 3600-turn coil for the primary and a home-made secondary of 90 to 100 turns. Alternatively, where the 12000-turn coil is available, use that as the primary with a 300-turn secondary.

For details of the oscilloscopes, see Appendix in *Teachers' Guide Year 4*.

For a class experiment, ask the pupils to set the volt/cm switch of the class oscilloscope to 1 ms/cm and the a.c./d.c. switch to a.c. Connect 2 or 4 V a.c. to the input terminals of the oscilloscope. Turn the variable control of the time-base anti-clockwise until four of five cycles of the wave form appear on the screen. When the CRO is correctly adjusted, the pattern traced on the screen will remain fixed in position.

Describing and averaging As pupils see the wave form of mains a.c. again, discuss the things we measure:

FREQUENCY, the number of complete cycles the current, or voltage, goes through in each second. Give the modern unit, hertz, as a short name for cycle per second – just as volt is the short name for joule per coulomb. Mention the frequency of the a.c. mains – and all supplies from transformers running from the mains.

PEAK VALUES Point to the peak values on the oscilloscope trace. Since the current or voltage only reaches a peak value at two instants in a cycle, and is far less for the rest of the time, we could hardly use the peak as a fair measure. Then what average can we use for measurements?

PLAIN AVERAGE The simple, arithmetical, time average over any whole number of cycles is zero. That is hardly a useful measure of real currents that heat things and run motors.

ROOT-MEAN-SQUARE AVERAGE (R.M.S.) We choose for measurements the type of average that fits measurements of heating – by a.c. just as by d.c. Since the rate of producing heat when a current I flows through a resistor R is I^2R, we average I^2 and take the square root to find the r.m.s. value.

Show a very valuable demonstration:

Demonstration 48
Comparison of r.m.s. and peak value

Apparatus

a.c. supply, 2 V	item 27 or 104
2·5 V MES lamp	92R
MES holder for lamp	92T
2 dry cells	52B
rheostat	52F or 541/1
oscilloscope	158 or 64

Procedure

Connect the 2 V a.c. supply to the lamp. Attach a crocodile clip to one lead, to act as a home-made two-way switch.

Also connect two dry cells and the rheostat to the lamp, the two way switch giving a choice between the two supplies. (A circuit board may be useful here for holding components.) With that

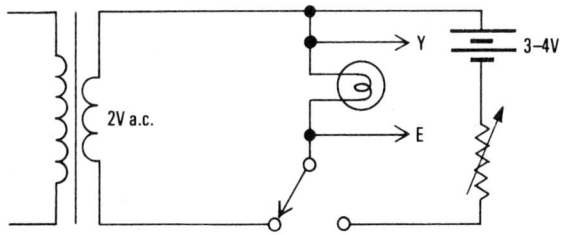

d.c. supply, adjust the rheostat so that the lamp glows with the *same* brightness as when on a.c.

Take leads also from the lamp to the input of the CRO with the time base running.

Switch to and fro between the two supplies, to make the comparison. With d.c., pupils see the trace deflected upward (or down). With a.c. they see the wave form with the peaks higher than the d.c. deflection.

(This is an important *qualitative* demonstration: the factor $\sqrt{2}$ for (peak value)/(r.m.s. value) could not be measured here.)

Meters for a.c. Tell pupils that there are ammeters and voltmeters designed to measure r.m.s. values. Nowadays these are usually d.c. (moving coil) instruments with a rectifier built in and a suitably labelled scale.

Ohm's Law with a.c. As a quick demonstration, let an alternating current from a low-voltage supply pass through a resistor and make enough measurements of p.d. and current to show that the ratio of the (r.m.s.) meter readings is constant. As in many demonstrations, a few measurements illuminate, but a long series fails because it is boring. Here, (0,0) and three other pairs will suffice.

Demonstration 49
Ohm's Law with a.c.

Apparatus

1 L.T. variable voltage a.c. supply	item 59
1 rheostat (10–15 Ω), as a specimen	541/1
2 demonstration meters	70
1 a.c. dial: 1 A	71/8
1 a.c. dial: 15 V	71/6

87

Procedure

Set up a simple series circuit using the a.c. output of the variable voltage supply, the demonstration ammeter and a rheostat (used here as a fixed resistor). Connect the voltmeter in parallel with the resistor. Adjust the voltage to make the maximum current about 1 A.

Record 3 or 4 pairs of measurements of current and p.d. as the voltage is gradually reduced.

Discuss the record with pupils. Do current and voltage go up and down together in proportion? Plot a graph.

one or two hertz, something with an obvious mechanism is required for our purpose. Our hand-driven device (due to Professor Pohl) seems best.

The generator A coil of high-resistance alloy wire is wound on a rectangular card and the ends are connected to a steady d.c. supply, 2 or 4 V. A revolving arm carries two metal brushes, 180° apart. The brushes rotate in contact with the coil and leads from them are brought out to provide the slow a.c. Three of the sketches (I, II, and III) show the brushes and the output in different positions.

The arm is turned by hand.

The generator works well. If it is dirty, a few drops of thin oil on the brushes will ensure continuous contact.

Details of the slow a.c. generator

Although simple transistor oscillators can easily be made to give an alternating supply of frequency

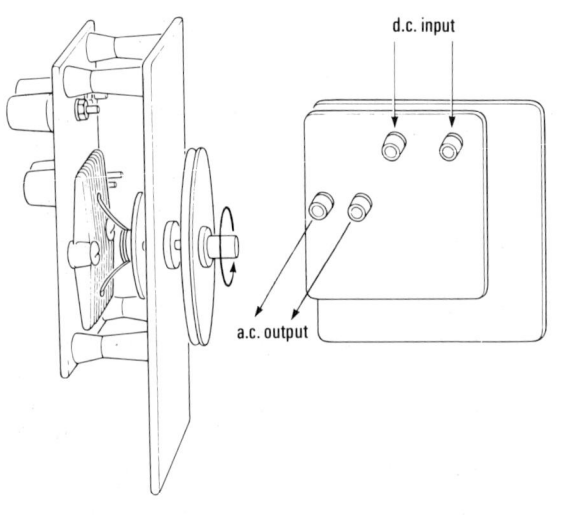

Class Experiment 50
Slow a.c. with low frequency generator

Apparatus

8 class oscilloscopes	item 158
8 low frequency a.c. generators	170
4 batteries (each to serve 2 generators)	176
8 resistors, 1 kΩ, $\frac{1}{2}$ W	132H
8 galvanometers	180

Pupils should work in groups of four.

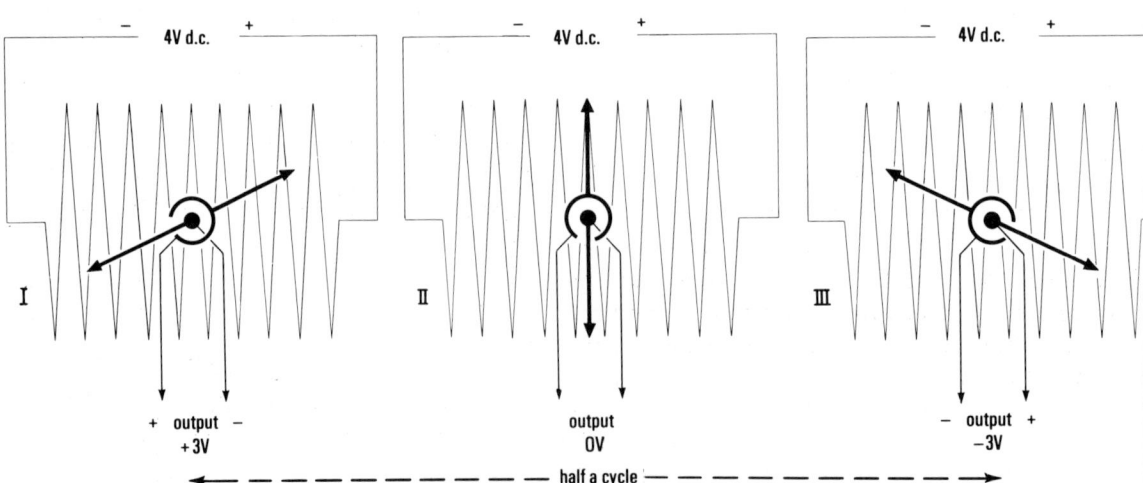

Procedure

Pupils follow these instructions.

    ★   ★   ★   ★   ★

a. *An alternating current* Try the output of the generator on a milliammeter (galvanometer). Connect the output terminals to the milliammeter with a 1000 ohm resistor in series to pass a suitable current. Turn the handle slowly.

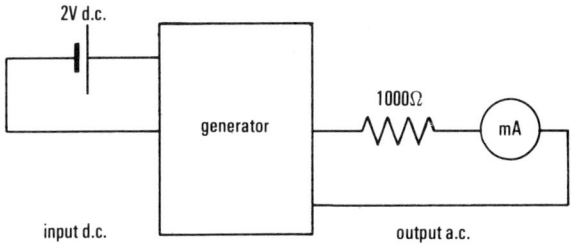

b. *An alternating voltage* Set the a.c.-d.c. switch on the oscilloscope to d.c.: turn off the time base and set the Y-gain to about 1.

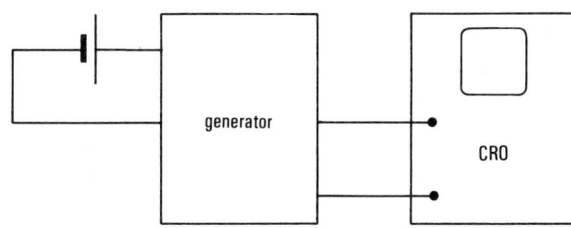

    Connect the output of the generator to the input of the oscilloscope. (It does not matter whether you keep the resistor in or not.)

    Turn the generator handle slowly and watch the oscilloscope. Then switch the time base to its lowest speed on range 1 as you turn the generator handle.

    Increase the time base speed and turn the generator faster.

    ★   ★   ★   ★   ★

    After the pupils have seen slow a.c. for themselves on the class oscilloscopes, teachers may wish to show the same thing with a demonstration meter.

Demonstration 50X
Slow a.c. with low frequency generator

Apparatus

1 low-frequency a.c. generator	item 170	
1 battery	176	
1 demonstration meter	70	
1 d.c. dial: 5 V	71/3	
1 a.c. dial: 5 V	71/5	

Procedure

Use the demonstration meter with a 5 V d.c. dial *and the pointer set centrally.* (Ideally a 2·5–0–2·5 V dial should be used, but this is not essential.)

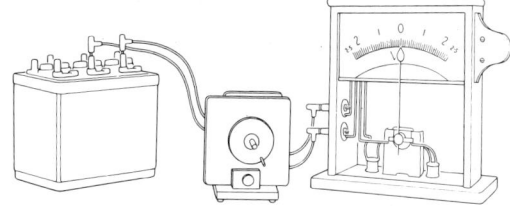

    Turn the generator slowly. Pupils watch the pointer moving back and forth.

    Then turn the generator faster and faster, so that the amplitude of the pointer's movement gets less and less.

    Then replace the d.c. dial with an a.c. dial.

Further experiments

The motor from the energy conversion kit (item 9A) may be used to drive the slow a.c. generator with an elastic band as the driving belt. The L.T. variable voltage supply (item 59) can be used to drive the motor, first at low and then at high speed. The output from the generator, driven by the motor, can be shown on an oscilloscope.

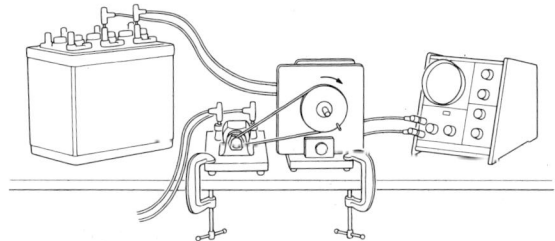

Class Experiment 51
Slow a.c. and a resistor

Apparatus

8 low frequency a.c. generators	item 170
8 class oscilloscopes	158
8 resistors, 1 kΩ, $\frac{1}{2}$ W	132H
8 galvanometers (centre zero)	180
4 batteries (each one to serve 2 generators)	176

The resistor and voltage have been chosen to give a good deflection when a centre-zero galvanometer 3·5–0–3·5 mA, is used. Other galvanometers may need different component values. 4 mm crocodile clip leads (Year 2 circuit board) are useful in connecting to the wire-ended resistance.

Procedure

Pupils follow these instructions.

*   *   *   *   *

Connect 2 V d.c. to the input of the generator.

Connect its output to the input terminals of the class oscilloscope which serves as a voltmeter. Connect the output terminals also to a 1000-ohm resistor in series with a centre-zero galvanometer.

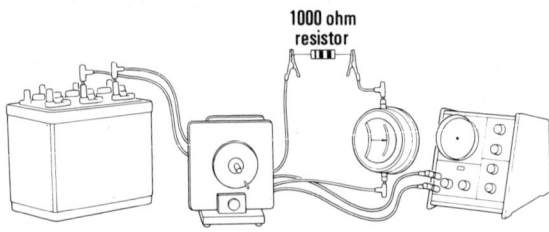

Keep the brilliance control of the oscilloscope as low as possible, and keep the spot out of focus, to avoid damaging the screen.

Start with the time base of the oscilloscope switched off and the a.c./d.c. switch in the d.c. position. Set the gain to 5 divisions per volt. Adjust the Y-shift so that the spot is in the centre of the screen when the galvanometer reads zero.

Turn the generator by hand at speeds of $\frac{1}{2}$ to 1 rev./second.

Watch and decide whether current and voltage

(p.d.) are in phase (that is, swing to and fro in step).

It is easier to compare phases if you hold the galvanometer with its face side by side with the oscilloscope screen and turn it so that the needle moves up and down like the spot.

If you have time, ask for a rectifier (from the circuit board kit) and insert it in one of the leads from the generator to the resistor.

*   *   *   *   *

It is advisable to visit each group of pupils in turn and make sure that they have connected their 'ammeter' and 'voltmeter' with the same polarity. With the rotating contacts stationary, the device supplies d.c.; then the galvanometer needle and the oscilloscope spot should both have moved up or both have moved down.

As before, teachers may wish to show this experiment again as a demonstration.

Demonstration 52
Slow a.c. and a resistor

Apparatus

1 low frequency a.c. generator	item 170
1 battery	176
2 demonstration meters	70
1 d.c. dial : 5 V	71/3
1 d.c. dial : 2·5–0–2·5 mA	71/4
1 resistor, 1 kΩ, $\frac{1}{2}$ W	132H

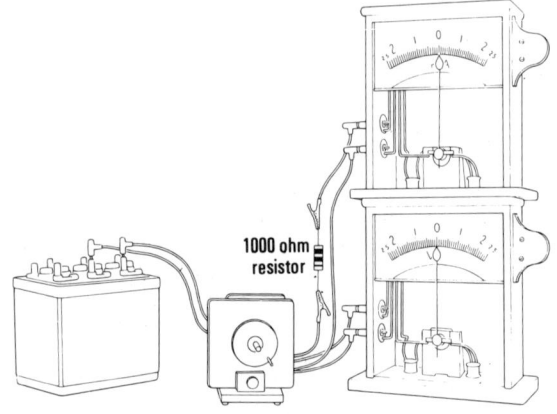

Procedure

Show the relationship between current and p.d. for a resistor with a.c.

Supply the input of the low frequency a.c. generator at 2 V. Set the pointer of the meter centrally (a dial for 2·5–0–2·5 V would be ideal, but it is not essential).

Connect the generator's output to the demonstration meter with the 5-V d.c. dial.

Also connect the generator's output in series with the 2·5–0–2·5 mA meter and a resistor of 1000 Ω. Crocodile clips can be used for connecting the resistor. (Connect the meters with the same polarity.)

Turn the generator by hand, at about 1 turn per 5 seconds. Pupils observe the voltage and current. They watch to see whether they are in phase.

Show what happens when the generator is speeded up a little.

Pupils continue with slow a.c. Although the first two class experiments with the low frequency generator will help pupils to understand alternating currents, they may not prove very impressive. Pupils will not be surprised when they try a resistor and find current and voltage across it changing in unison. So it would be a pity to let their class experiments end like that. Let them go on to a trial with a *capacitor*. Then the behaviour will be novel.

Class Experiment 53
Slow a.c. and a capacitor

Apparatus

8 low frequency a.c. generators	item 170
8 class oscilloscopes	158
8 galvanometers (centre zero)	180
4 batteries (each to serve 2 generators)	176
8 electrolytic capacitors (500 μF, 50 V working)	132C

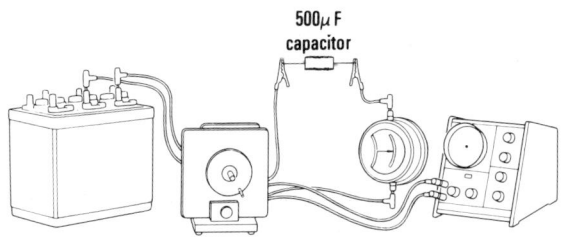

500μF capacitor

In a class experiment 100 μF will prove disappointing. We trust schools will be able to provide enough 500 μF capacitors.

Preparation

Electrolytic capacitors should normally be used only on direct voltages with the correct polarity. However, it has been found that they work well in these experiments with very low frequency a.c. What matters is the allowable ripple current. Lists from suppliers will show that the currents in this experiment are much smaller than the allowable ripple current.

The lack of any direct polarizing voltage may cause some deterioration of the dielectric and it is a wise precaution to 'form' the plates before and after use. This is done by connecting the capacitors to d.c. with a voltage less than or equal to the working voltage of the capacitor, and of the correct polarity.

Procedure

Pupils read a description in *Pupils' Text*, then follow these instructions.

    ★   ★   ★   ★   ★

This experiment is just like the last one except that you replace the resistor with a capacitor.

You should follow the same instructions as for **Experiment 51**.

As you turn the generator by hand, at about 1 rev./second, watch the galvanometer. The pointer will move, but not very far. For greater motion you would need more voltage or a larger capacitor – one with a larger area of plates. Consult your teacher about using a larger d.c. voltage on the generator.

Now compare voltage and current as you did for the resistor. This is easier if you hold the galvanometer on its side. Both voltage (on the oscilloscope) and current (on the galvanometer) change as the generator produces the slowly alternating voltage. But do they keep in step? Are they in phase? Does the current reach maximum at the same instant as the voltage?

Discuss that behaviour of current and voltage with your teacher.

    ★   ★   ★   ★   ★

Thus a capacitor appears to let current through, but the current is out of phase with the driving voltage by $\frac{1}{4}$ cycle – as the experiment shows.

In terms of pupils' understanding, the personal experience of the surprising result with the capacitor is most fruitful. Teachers may however wish to supplement the class experiment with a demonstration using two meters.

Demonstration 54
Slow a.c. with a capacitor

Apparatus

1 low frequence a.c. generator	item 170
1 battery	176
2 demonstration meters	70
1 d.c. dial: 5 V	71/3
1 d.c. dial: 2·5–0–2·5 mA	71/4
1 electrolytic capacitor (500 μF, 50 V working)	132C

Preparation

See the comments on electrolytic capacitors in Class Experiment 53.

Procedure

As in the Class Experiment, except that the oscilloscope is replaced by the demonstration voltmeter with its pointer set to the centre position for zero voltage.

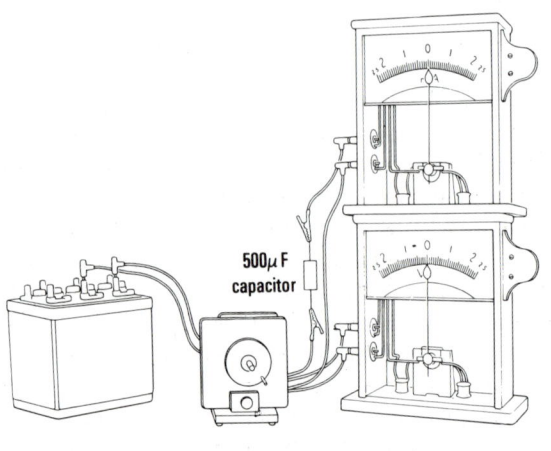

500μF capacitor

DISCUSSION

Charges and current? In the case of a capacitor, we may talk about positive and negative charges of electricity surging on to one plate and off the other, then back again to zero, then the same in reverse – driven thus by the alternating voltage.

There is a good insulator between the plates, so the current (in the ordinary sense) cannot pass through the capacitor – it only appears to do so. But if we put a lamp in the circuit, the charges go through the lamp on their way to and from the capacitor plates and the lamp lights. A quick demonstration will help.

Demonstration 55
Capacitor on a.c. with a lamp

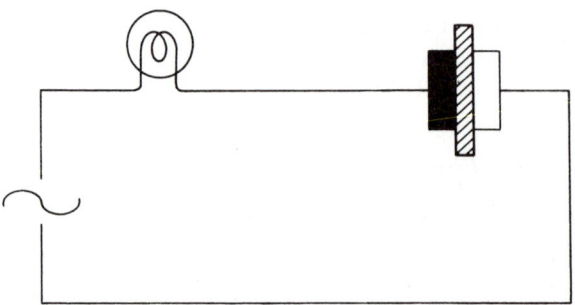

A large capacitance is needed, so an electrolytic capacitor should be used. Although normally intended for use with d.c. of correct polarity, an electrolytic capacitor will operate on a.c. but within limits. What matters is the allowable ripple current. So, if we wish to use a 240-V, 60-W lamp, we need a capacitor which can accept a ripple of at least 250 mA. One might expect success with a capacitor rated about 50 μF, 450 V working but suppliers' lists should be consulted for precise details. The plates should be formed just before use with a d.c. voltage slightly below the working voltage.

Model Transmission Line with Alternating Current

All pupils should now try the model transmission line again, this time with a low voltage a.c. supply. Using only that supply, they find it as inefficient as with a similar d.c. supply. Then, *in a demonstration*, step-up and step-down transformers are inserted to provide a large voltage between the power-line wires.

Class Experiment and Demonstration 56
Power or transmission lines

Apparatus

8 pairs power line terminal rods	item	99
16 1¼-m lengths covered Eureka wire (diameter 0.40 mm)		98
16 retort stands and bosses		503–505
16 SBS lampholders on bases		74
16 lamps (12 V, 24 W)		72
4 (or more) 12-V batteries		176
8 transformers		27
2 pairs C-cores and clips		92G
2 coils (120 turns) to fit C-cores)		127
2 coils (2400 turns)		128

Pupils work in groups of four.

Preparation

It will save much time and avoid confusion if the pylons and power lines can be set up beforehand. Otherwise assembling the apparatus may bulk larger in later memory than the very important exhibit of power transmission and the need for high voltage.

For each group of pupils, two dowels form the power-line terminal rods. They are held horizontally in bosses at a height of 30 to 50 cm above the bench and 1 metre or more apart. Two lengths of high resistance wire (Eureka SWG 28) are stretched between the terminals to form the power line.

Even though most pupils will have tried the low voltage d.c. power line in an earlier Year, all should start with it. It will enable them to get the apparatus in order and review its working.

Procedure: Class Experiments a, b

a. Low voltage d.c. power line

Pupils follow these instructions.

*   *   *   *   *

Set up the experiment sketched. Use a 12-V battery (or similar power supply) as your 'power

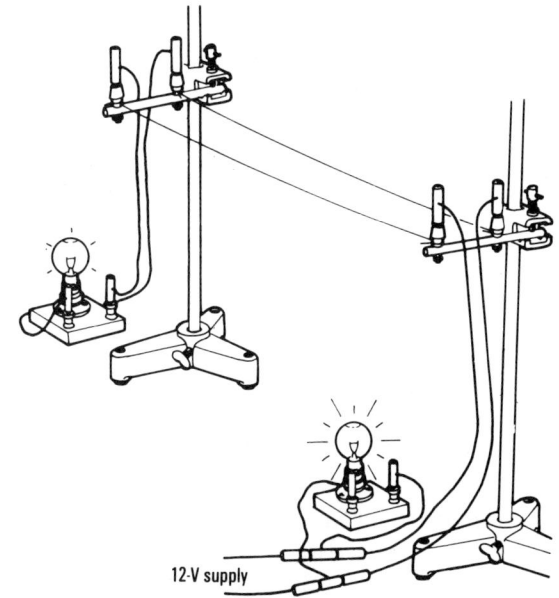

12-V supply

station'. Run wires from the 'power station' to the terminals on a pylon nearby.

Run two thin wires from the pylon to a second pylon at a 'village' far away. Those wires are your model power line.

At the 'village', connect a 12-V lamp to the power line.

Switch on. How well is the 'village' lit?

b. Low voltage a.c. power line

Keep the same arrangement as in **a** but use the 12 V a.c. terminals of your transformer for the 'power station'. How well is the village lit? (You will not need to use an ammeter or voltmeter.)

*   *   *   *   *

c. Demonstration power line with high voltage a.c.

Take a power line assembly made with covered Eureka wire of the same gauge and show the advantage of using high voltage between the transmission lines. The experiment *must* be done as a demonstration.

Use the same low voltage a.c. supply for the 'power station' but install a step-up transformer between the 'power station' and the line, and a step-down transformer between the other end of the line and the 'village'.

For each of these transformers use a pair of C-cores and a clip with a 120-turn coil on one leg and a 2400-turn coil on the other.

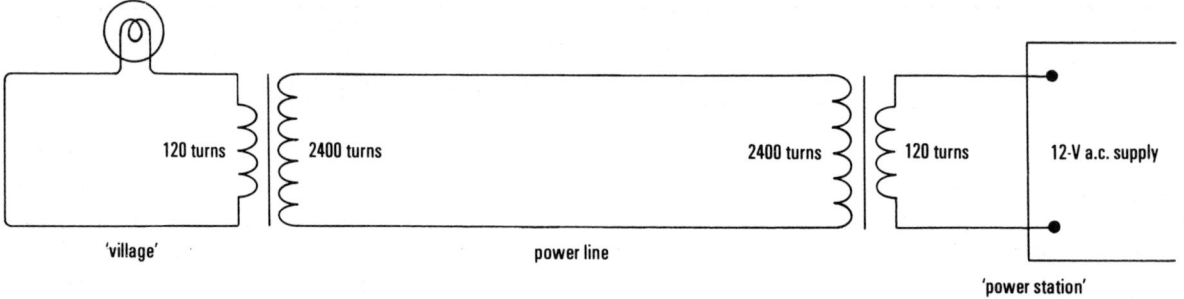

'village' power line 'power station'

120 turns 2400 turns 2400 turns 120 turns 12-V a.c. supply

It is important to use exactly the same gauge of covered wire and *low* voltage lamps in this demonstration as in pupils' own experiment. But now there is not the same loss in the power line.

A voltmeter (or perhaps a 240 V lamp) might be connected across the two wires of the power line to indicate the high voltage there.

Note. Remember that there will now be a p.d. of two or three hundred volts between the two wires of the power line.

A power grid Discuss the advantages of a.c. over d.c. for a power grid. Transformers have high efficiency and need little supervision, while motor-generators for transforming d.c. voltages need watchful engineers.

Mention the disadvantages: alternating voltages lead to losses in underground cables. With cables under the sea (water conducts very well) the losses would be too great. In exchanging electric power between Britain and France we have to use d.c. under the Channel, with costly converters at each end.

CHAPTER 8

Waves and theories of light

The work of this chapter needs to be kept to a very short time. If it takes long, it will defeat its own objective, because its aim is preparation for atomic models in Chapter 11. If this preparation takes a long time, pupils may never reach wave models in Chapter 11! This chapter offers:

Reminders of Year 3 topics:

Experiments for catching up (for pupils who missed the Year 3 work or did not understand it)

Examples of waves (including some new ones)

Interference: Young's fringes in a ripple tank★
Young's fringes with light★

Diffraction: Ripples passing through gaps★
Examples with light★

Theories: waves or particles for light★

[★ These experiments were marked *OPTIONAL NOW* in Year 3. They are needed in Year 5. See *General Introduction*, page 40, for comment.]

New Topics:

Standing waves; qualitative examples

Young's fringes; estimate of wave length of light.

The *immediate* aims are to remind pupils of the way waves behave and to clarify ideas of speed, wavelength, and frequency; to remind pupils of interference of light (Young's fringes); to compare theories; to prepare for diffraction gratings in Chapter 9; and to show examples of standing waves – treated qualitatively as modes of vibration.

The *ultimate* aim is preparation for atomic models, mentioned above. With that in mind, waves must be treated, but intensive teaching would not be justified.

{**Newcomers and old hands** For this chapter we might consider pupils in three groups, divided according to past work in physics – quite independently of differences of ability:}
{(A) Newcomers who missed Nuffield Year 3 and never worked with ripple tanks or made Young's slits for interference.}

{(B) Old hands who did many an experiment with their own ripple tank in Year 3 but missed the experiments marked *'OPTIONAL NOW'* on diffraction and interference of ripples and/or those on diffraction and interference of light and/or the comparison of theories of light.}
{(C) Old hands who did all the diffraction and interference experiments with ripples and with light in Year 3, and compared theories of light.}

{Pupils in Group (A) – if there are any – should either omit the rest of this chapter and most of Chapter 9 or be given special help, if possible in the form of quick class experiments rather than demonstrations. Members of Group (B) could both learn for themselves and be valuable teachers if they could be paired with members of Group (A). Think how quickly they could get members of (A) through the housekeeping work of ripple tanks!}

{Apart from acting as teachers, members of Group (B) should be urged to proceed straight to the experiments with ripple tanks and with light that were marked *OPTIONAL NOW* in Year 3, but are necessary in Year 5.}

{Members of Group (C) might move to *Demonstrations for Revision* until they meet the one experiment that will be new to them: measurement of wavelength of light by Young's fringes – good preparation for gratings in the next chapter.}

WAVES

Revision? Wave motion should have been studied qualitatively early in Year 3. In extending those studies now we should remember that many pupils will not proceed to further physics after this Year. So we should *not* make an extensive revision of the earlier work. We should only do what seems interesting and relevant.

Wave models or examples? Mechanical wave models are expensive and often seem to pupils rather 'special' – gadgets specially designed to imitate waves. Yet waves are general phenomena. We suggest it is more important to show some examples of real waves in common media, rather than well built models.

So we do *not* suggest that schools should buy models. Instead, demonstrate some real examples.

---

†Demonstration 57
Examples of wave motion

Apparatus

a.
1 length of rope★ (about 6 m)

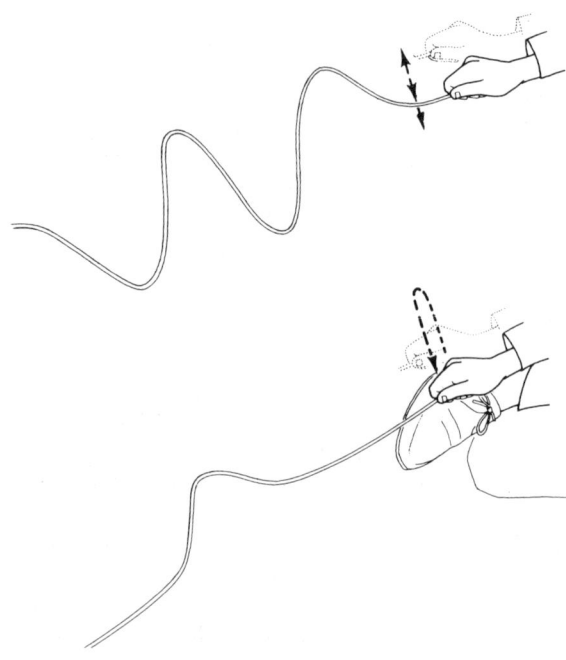

b.
1 long 'slinky' spring   item 101

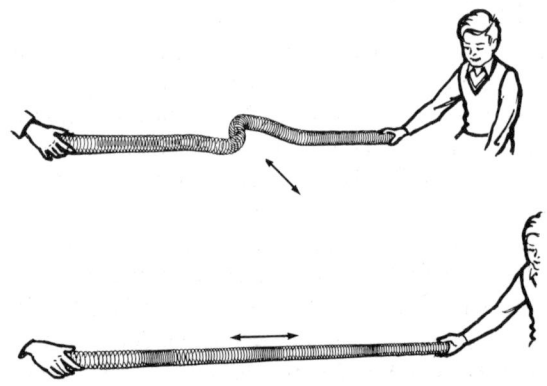

c.
1 rectangular tank† with wooden   item 100/2
   block or paddle

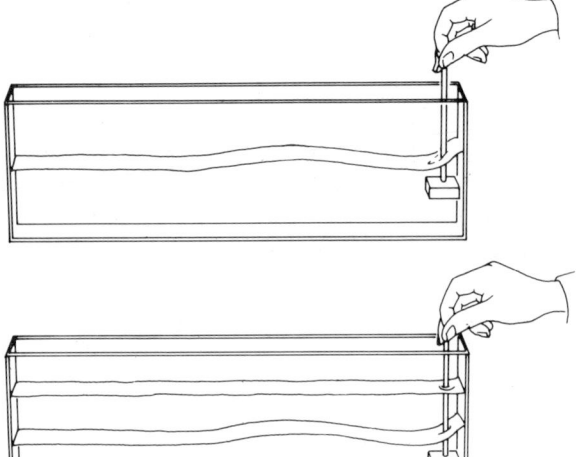

paraffin

★ The longer the rope the better, depending on available space. It must be flexible and fairly massive. Clothes line is too stiff; string is far too thin. A long rubber tube could be used instead.
† See the note about this tank in *Teacher's Guide Year 3*, page 8.

Procedure

a. *Waves on a rope*† Stretch the rope on the floor or bench top. Fix one end and move the other end at about 4 Hz.

It is best if the rope is long enough for the wave to die out considerably by the time it reaches the far end, so that reflections do not cause large standing waves.

Pupils watch a continuous train of waves travelling along the rope.

b. *Waves on a slinky*† Stretch the slinky across the floor, or along the bench top. First show a transverse pulse then a continuous transverse wave travelling along it.

Also show longitudinal pulses travelling along the slinky.

c. *Water waves seen from the side*† Half fill a long rectangular tank with water. Place it so that pupils can see the water line face on. Then they see any waves passing along in section.

Generate waves near one end by moving a hand or a block of wood up and down in the water. (Some teachers prefer to do this by sweeping to and fro with a wooden paddle.)

If a little sawdust is mixed in the water, pupils who are very close to the tank may see the path of individual particles in the medium – vertical circles for particles at the surface, ellipses for those deeper down.

A better arrangement, which shows much slower waves, uses the interface between two liquids. Fill the tank about one-third full of water and add paraffin (preferably coloured) above that until the tank is two-thirds full.

Generate transverse waves at the interface, keeping the block or paddle immersed.

Demonstration 58
New examples of wave motion

These were not suggested in Year 3. They are not essential now but are good embellishments.

Apparatus

a.

1 Edinburgh CO_2 pucks kit	item 95
4 extra CO_2 pucks	169
magnetic strip, 2 m	95G
window-cleaning liquid (or methylated spirits) for cleaning plate	
1 CO_2 cylinder	19/1
1 dry ice attachment and cloth	19/2

b.

20 dynamics trolleys	106/1
40 expendable springs	106/1
40 expendable springs	2A
60 trolley pegs (dowels)	
1 10-cm G-clamp	44/1

a. *Waves with pucks* Place dry-ice pucks (ring magnets) in a line, short equal distances apart, on the levelled glass sheet. Instal a fence of magnetic strip on each side of the line.

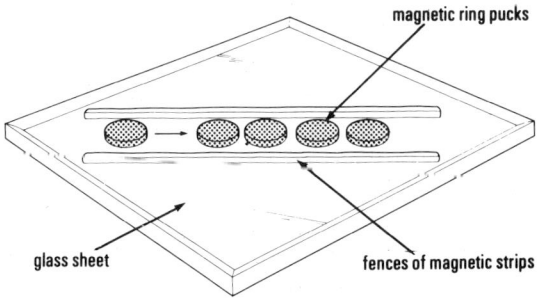

magnetic ring pucks

glass sheet   fences of magnetic strips

Pull the end puck a little way back then give it a push towards the others: pupils will see a compression wave.

b. *Waves with trolleys* (*i*) *Longitudinal wave* Arrange a line of trolleys as in sketch I.
(*ii*) *Transverse wave* Arrange trolleys as in sketch II. Clamp one trolley to the end of a smooth bench. Connect the others with springs as in the sketch.

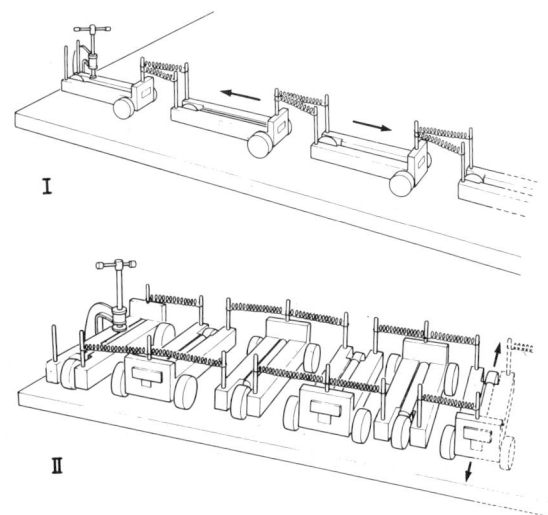

I

II

Keep the trolleys a short distance apart so that they can roll to and fro without hitting each other.

Give the trolley at the other end a sudden deflection to one side and back, to start a transverse pulse.

Then make the end trolley oscillate continuously to show a train of waves.

Unless the bench is very rough, or covered with cloth (velvet), reflections from the clamped end are likely to cause standing waves.

SOUND WAVES

{Many teachers will wish to spend some time on sound and music. This is an admirable part of physics for good clear teaching with some fine demonstrations. It is very interesting to some pupils, but dull and irritating for some less musical ones. So we do *not* suggest treating sound in this programme – except for a brief mention of sound waves – simply because we feel the time is needed for other parts of physics, especially atomic physics. There are already good books for pupils with a special interest in this field.}

STATIONARY WAVES

Pupils should see examples of standing waves. We urge teachers to avoid the difficult idea of standing waves being produced by two trains of moving waves. (That is an artificial way of treating what seems clearly to beginners a simple vibration pattern, not a wave. If we insisted on the wave-synthesis story, we should make what is simple seem complex and artificial.)

Pupils need to gain a strong physical picture of standing waves if they are to use the concept in atom models.

Demonstration 59
Standing waves

Apparatus

6 metres of rope (soft, flexible)
1 long slinky   item 101
1 large rectangular transparent tank   100/2

Procedure

a. Tie one end of the rope securely to a fixture on a wall. Pull the other end taut. Move that end up and down to excite transverse waves. Build up a pattern of standing waves by feeling for the right resonant frequency and adjusting the tension.

A more effective method is to drive the motion at a *node*. Secure the rope firmly at *both* ends. An ink mark or tape at, say, one-fifth of the length from one end marks the best place to drive the rope with finger and thumb held loosely in a ring. Move the hand up and down with the rope loose in the ring and change the frequency until the 5-loop motion builds up.

b. Build up a longitudinal standing wave on a slinky. Clamp both ends of the well stretched string and excite it by hand near a node.

c. (*Optional*) Show standing waves in the transparent tank half full of water while pupils view the water-line from the side.

We suggest a monochord demonstration as a good additional example of a standing wave. It is of interest to musicians and it helps to make standing waves seem real.

Demonstration 60
Music from standing waves: monochord

Apparatus

1 monochord†
1 tuning fork and rubber hammer
1 violin bow
paper 'riders'‡

STANDING WAVES ON A STRING

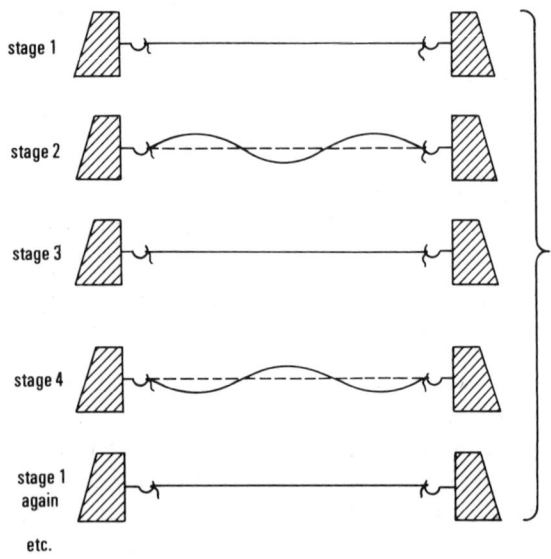

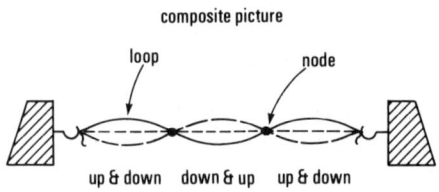

composite picture

loop   node

up & down   down & up   up & down

† The monochord need not be an elaborate instrument with a sounding box on legs. A wooden board, 1 metre or more long, with a steel piano wire strung between two nails or screws will do well. (There must be a board, to radiate sound. A wire stretched between immovable anchorages would radiate almost no sound.) Two triangular bridges are needed, to limit the vibrating length. See sketch I.
‡ To make the 'riders', take a strip of paper, say 3 cm by 10 cm, fold it in half along the middle line, and chop $\frac{1}{4}$-cm V-shaped pieces from it, as in sketch II.

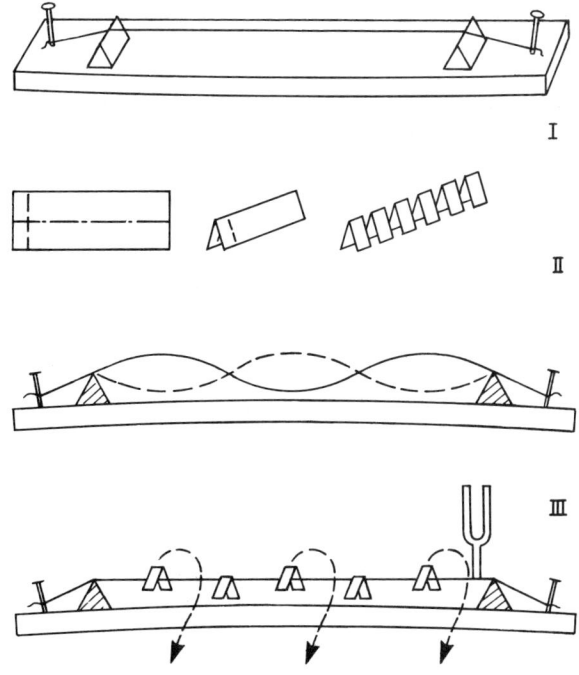

Musical instruments Tell pupils that violins, 'cellos, etc., are played with the basic note, the string vibrating in one loop – two or more loops only as a special trick. Yet the motion of the bowed or plucked string is seldom *simple* harmonic. It is usually 'compound harmonic', a mixture of the notes of motion in 1 loop, 2 loops, 3 loops, and so on.

The air in pipes – organ pipes, flutes, bugles, whistles – can vibrate along the pipe in the equivalent of 1 loop, 3 loops, 5 loops, etc., and sometimes 2 and 4 loops. An organ pipe or a clarinet gives 'compound harmonic motion' but the notes with several loops alone are not played unless the pipe is overblown. With a bugle, however, the different available notes are just those with air motion in several different numbers of loops. (The basic note, equivalent to one loop, is ugly and is not played.)

WAVE TRAINS: SPEED, FREQUENCY AND WAVELENGTH

Pupils' Text defines, or rather, describes wavelength and frequency and arrives at a relation connecting speed with them.

The 'wave formula' Since any measurements that might be done with ripples on water in a ripple tank are necessarily rough it is probably better to tell the pupils that $v = fL$. We might say as in *Pupils' Text*:

Suppose a man marching along takes 80 strides a minute, each stride $\frac{1}{2}$ metre long. How far does he go in a minute? . . . Yes, 40 metres.

His speed = [80 strides per minute]
$$\times [\tfrac{1}{2}\,\text{m per stride}]$$
$$= 40\,\text{m per minute.}$$

Suppose he takes f strides per minute, each of length L, in metres. Then he travels a distance fL metres in a minute.

His speed = [no. of strides per minute] × [stride]
or $v = fL$

Now think of waves. When a train of waves has travelled one wavelength along, it looks exactly the same again. One wavelength is like one stride. If the vibrating bar in the ripple tank turns out f whole wavelengths per second (frequency in hertz or cycles/second) the wave speed is given by:

Procedure

a. Stretch the wire taut and excite it by plucking or bowing it near one end. Pupils listen to the note.

Then pluck or bow it again, at the same time touching it *very lightly* with a finger at its midpoint. That will start the wire vibrating in two loops, with double frequency, an octave higher. Without telling them this, ask pupils how the note has changed.

Repeat with a finger touching lightly at $\frac{1}{3}, \frac{1}{4}$, etc. of the wire's length.

b. Show the existence of nodes and loops by hanging paper riders at appropriate places. Place the finger to suggest a suitable node, and bow gently, briefly. Riders at loops will jump off, those at nodes should stay there.

c. The procedure of (b) needs practice. An easier method is to excite the wire to resonance with a tuning fork. Tune the wire beforehand, by changing the length between bridges, so that it vibrates in, say, three loops with exactly the frequency of the fork.

Let pupils listen to the fork and to the note of the wire in the chosen harmonic. Place riders on the wire and excite it by touching the wire at one bridge with the shank of the vibrating fork.

WAVE SPEED = [NO. OF WAVELENGTHS
          TURNED OUT PER SECOND]
          × [WAVELENGTH]
WAVE SPEED = [FREQUENCY] × [WAVELENGTH]
$$v = fL$$

We should not drive home the result by asking for it to be written down in formal fashion and learnt, or by giving a long series of examples on it. At this stage we merely mention it as a case of a relationship between some measurable things. We might say that there are many relationships in physics like that, which are interesting and useful parts of our knowledge.

Some pupils may raise questions about radio waves. Do the BBC statements of wavelength refer to the same thing as our wavelength here? Does the BBC tell us frequencies? What are kilohertz, megahertz, etc.? Having answered those questions, we can use data from the BBC to calculate the speed of radio waves. Of course we must not use this as a scientific way of discovering the speed of electromagnetic waves. We must make it quite clear that we are only working out a number to tell us what the BBC have already assumed is the speed of its waves.

THEORIES OF LIGHT

In Year 3 there was a section of 'Continuation Experiments' (marked 'OPTIONAL NOW'). These dealt with rival theories of light. Unless pupils followed that discussion in Year 3, we should compare theories now. The sign † indicates material from Year 3.

† A bullet theory for reflection? Ask whether reflection of a ray of light and the way in which light travels in straight lines both fit well with the idea that light is a stream of speedy bullets.

If necessary, bounce a ball.

†Class Experiment 61
'Reflection' of a particle

Apparatus

Rubber ball(s)

Procedure

The path of an ordinary rubber ball (tennis ball, squash ball, or toy) bounced obliquely on a wall or floor will show something like equality of angles.

Pupils should try to compare (roughly) the angles the ball's path makes with the wall before and after collision. However, this is difficult in a demonstration: personal experience is better. Fortunately, many pupils know the answer already: others should be asked to go and try.

A superball is unsuitable. It is apt to suffer strange changes of motion! A spinning or rolling ball is unsuitable. Otherwise it would be simple for each pupil to roll a marble along the table to hit a glass block obliquely; but the rolling motion upsets the angle relationship.

Refraction Remind pupils of the way a ray of light is bent as it passes from air to water. If necessary, demonstrate it (*Teachers' Guide Year 3*, Experiment 33, page 96).

Now in Year 5, *Pupils' Text* argues that a 'bullet' of light would have to be *attracted* as it enters water and therefore would *move faster in water*.

Let pupils try a qualitative model.

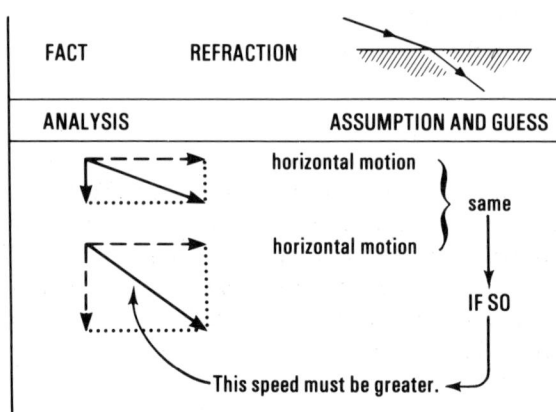

†Class Experiment 62
Particle model of refraction

Apparatus

1 kit for particle model of refraction item 96
Contents
    8 hinged hardboard platforms   96A
    8 launching ramps   96B
    8 steel balls (22 mm)   96C

This should be a quick, qualitative experiment, just to watch what happens.

The two pieces of hardboard are hinged underneath. Pupils raise the larger piece on a wooden block or books.

Procedure

Pupils follow these instructions.

*   *   *   *   *

Set up a small level plate to represent 'air', with a short downhill slope to the table. Gravity will make its successful tug on that slope. Arrange a

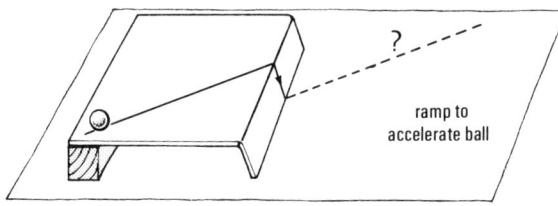

ramp to
accelerate ball

launching ramp for the ball on the upper plate. Let the ball roll along a slanting path. Watch what happens to its path, and its speed.

This model suggests that light bullets (*if* that is what light is) must move faster in water or in glass than in air.

*   *   *   *   *

Fast pupils may also try launching the ball on the lower surface to see 'refraction away from the normal' as the ball slows on going up the ramp. For this the height of the upper board must be reduced to about $\frac{1}{2}$ cm. It is also possible to see 'total internal reflection'.

Discussion Nearly three centuries ago, Newton speculated about light. He decided in favour of a bullet theory, because he saw that it would account for straight rays and sharp shadows. Meanwhile Huygens in Holland developed a wave theory, but it was not until much later (about 1800) that waves of light gained strong experimental support, from the work of Thomas Young.

{Thus – as Newton knew – a bullet theory accounts for the behaviour of light quite well, and makes straight rays of common (non-laser) light and sharp shadows seem much more reasonable than a wave theory would.}

{There were two serious difficulties ahead:
(*i*) The fact that *some* of the light is reflected at a boundary and *some* of it is refracted. How can bullets split their work like that? (It is probably wiser not to raise (*i*) with pupils.
(*ii*) The phenomena of interference and diffraction.}

{The two difficulties together forced Newton to suggest a strange scheme that endowed the interface with alternating 'fits' of easy reflection and easy transmission. He knew very well that his theory implied some periodic activity connected with the moving bullets, so that one could assign a wavelength – yet he did not change to a wave theory, because he considered sharp shadows too difficult to reconcile with waves.}

{Long after Newton's time, the speed of light in water was measured and compared with the speed in air. Light travels *slower* in water.}

{That is generally said to be a *crucial* experiment which decided clearly against the bullet theory. Yet most crucial experiments, if not all, are only crucial – leading to an inescapable decision – if one sticks to the full details of the theories being tested. Newton's prediction assumed that the MASS of a light-bullet remains constant, and that its component of MOMENTUM along the interface is conserved – by symmetry and Newton's Law II. On the basis of constant mass, that predicts greater speed in water. If we allow the bullet to change its MASS but conserve its total KINETIC ENERGY, the prediction is reversed; smaller speed in water! Thus not even here is there a fully crucial test unless we choose a particular set of assumptions.}

Refraction of waves If pupils looked at refraction of ripples in their class experiments with ripple tanks in Year 3, ask what happened, and if necessary bring out a tank for a quick look. If they postponed that part of the ripple-tank series, they should try it now. Although it is difficult, personal trial in groups of four *with plenty of help* will give the sense of genuine experience that is needed.

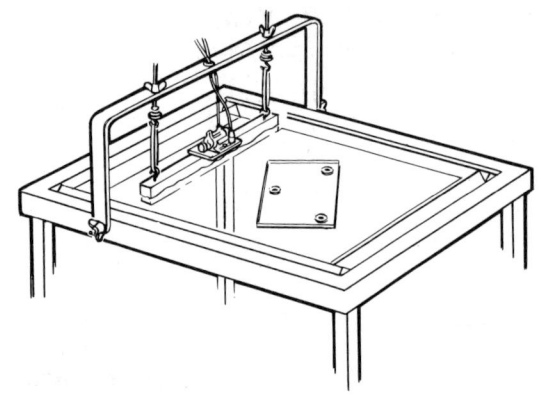

Class Experiment 63
Refraction of ripples

Apparatus

8 ripple tanks	item 90
8 motors mounted on beams	90L
8 plates of glass (*rectangular*)	90P
32 nuts or washers	
32 hand stroboscopes	105/1

The glass plate should be a rectangle. (The irregular shape often suggested is confusing.) The nuts or washers which act as spacers should be 3 or 4 mm thick. The water should be very clean. Oil or dirt increases the friction and makes the refracted waves even harder to see.

Procedure

Pupils follow these instructions.

★   ★   ★   ★   ★

Ripples travel at a different speed in shallower water. But they also died away soon, because their energy is taken away by water friction. However, if you level your tank and adjust things very carefully you should be able to see what happens.

To make a patch of shallow water, put a sheet of glass in the tank. (Put some small bits of metal under the glass so that you can take it out again easily.) Pour in water until it *just* covers the glass. Then take out a little water, leaving a *very* shallow layer of water above the glass. (The tank must be very carefully levelled for this.)

Lower the vibrating bar until it just touches the water. Run the motor *very* slowly.

(*i*) Arrange the glass sheet so that the ripples meet its edge head-on. What happens to their speed? What happens to their wavelength (distance from crest to crest)?

(*ii*) Turn the glass so that the waves will meet it in a slanting direction. Now watch carefully.

Make a sketch of what you see, showing how the *direction* of the waves changes when they meet the shallower water – and, as you know from (*i*)

they travel slower. Add to your sketch a 'ray' perpendicular to the waves, to show their *directions of travel*.

★   ★   ★   ★   ★

Discussion Recapitulate the results:
(*i*) Ripples travel slower in shallow water than in deep water.
(*ii*) If ripples strike the boundary between deep water and shallow at a slant, they are bent. Their 'guide-lines' of travel, or 'rays', bend.

Illustrate the second fact with a sketch. Then compare that sketch with the bending of rays of light when they pass from air to water.

Lead to the conclusion: *if* light consists of waves, the refraction of light shows that it must travel *slower* in water than in air – the opposite of the story for bullets.

Then we can ask for a 'crucial experiment'; a comparison of the speeds of light in air and water. Unfortunately light travels so fast that we cannot give a demonstration – as we could for sound, with an oscilloscope. We must simply tell pupils the result: slower in water, only $\frac{3}{4}$ as fast.

Waves versus bullets Pupils have seen that a bullet theory of light ('a thinking-model') fits with straight-line rays, sharp shadows, reflection, and (*if light travels faster in water and glass than in air*) with refraction.

But we have also told pupils that some scientists developed a wave theory; and we asked for a choice between the two views. At first that was just a matter for guessing or for accepting some assertion. But now wave theory predicts '*slower in water than in air*' and the 'crucial experiment' seems to decide in favour of waves.

{Yet we should announce the decision with a

102

little hesitation. A 'crucial experiment' provides a *fact* but its *interpretation* may be twisted.}

So we encourage pupils to see some more experiments before making up their minds – examples of diffraction and interference of light. And, in a later chapter, they see wave and bullet behaviour for both light and particles of matter.

Details of apparatus for diffraction and interference

VIEWING SCREEN. The screen at which pupils view diffraction patterns or interference fringes must be a translucent one that scatters light through a *small* angle. Each pupil in turn stands immediately behind the screen and receives the light that comes almost straight through it to him.

Kitchen greaseproof paper, waxed paper, oiled tissue, are all good for this. (So is translucent plastic sheet, but that is unnecessarily expensive.)

One might expect a screen that scatters the transmitted light through a *large* angle to enable other pupils standing at one side to view the pattern; but the pattern would then be far too faint. (Architects' tracing linen scatters through a *large* angle – that is what makes it so useful for an illuminated screen placed behind apparatus to show it in silhouette to a widespread audience in front. But that is not what we need. Here, tracing linen is unsuitable.)

LIGHT SOURCE FOR ALL DIFFRACTION DEMON-STRATIONS. The compact light source is essential for brightness. And its small filament area is near enough to a point source for *some* of the effects to be seen.

For clearer patterns, and especially for the shadow of a disk or ball, the source must be still smaller. Place a metal screen with a 1 mm hole in front of the compact light source, as close as possible. A lens is not necessary, and it would spoil the simplicity.

LIGHT SOURCES FOR CLASS EXPERIMENTS ON YOUNG'S FRINGES. Use ray-streak lamps, perhaps run at higher voltage. They must have very straight vertical filaments. 36-W lamps would be better than 24-W ones.

Surround each ray-streak lamp by its shield, with the long wide opening in the shield serving as doorway for the light beams. A few centimetres from the door-way, place a home-made 'limiting

screen' – tailored to fit the arrangement of the laboratory.

LIMITING SCREENS. Eight groups of pupils will need 8 beams of light traversing the room but it is not necessary to set up 8 lamps. One lamp can provide 3 or 4 beams if a limiting screen is set up in front of it.

Such screens are necessary to cut off stray light in the dark room. The visibility of the pattern on the distant translucent screen is easily spoiled by stray light or by light *reflected from shiny bench tops*.

If a lamp is to serve three groups, the limiting screen should be about $7\frac{1}{2}$ cm square with three coarse slits in it, to let out three beams of light. These can be rough slits cut in a piece of tin plate with snips. No precision is necessary – these are light shields, not combs for ray streaks. If the screen is placed 5 cm from the lamp filament the slits should each be 2 to 3 mm wide, spaced about 14 mm apart. Then each of the three beams will cover a slide with double slits 1 or $1\frac{1}{2}$ m away.

GLARE. Unless the lamps are high up, there is danger of light on its way across the room being reflected by polished table tops – reflection is copious at very oblique incidence. So black cloth should be laid on table tops, or extra screens placed to stop reflected light from reaching the viewing screens.

MATERIAL FOR DOUBLE SLITS. The slits are scratched in a soft black film on glass. Paint microscope slides quickly with Aquadag (colloidal graphite), using a soft brush. Allow the slides to dry.

Before the slits are ruled, hold each slide up to a bright light and make sure it is thoroughly opaque.

RULING THE SLITS. Each pupil should do this. There are several good ways of ruling slits on the coated slides, ranging from simple hand scribing to the use of a special device.★

1. *Rough ruling by hand.* Drag a blunt pin along a metal ruler. To make the second slit, hold the ruler there but tilt the scriber a little and drag it along

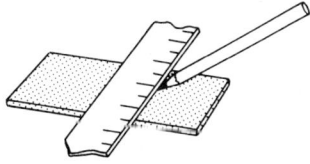

★ The ruling of double slits – like measurements of *g* and the monkey-and-hunter demonstration – will often attract the

again. A biro as a scriber is even better, but it must be the very fine kind. *This is the method that we recommend strongly for pupils.*

2. *The eye of a darning needle*, broken across, can be used as a miniature pitchfork to rule double slits.

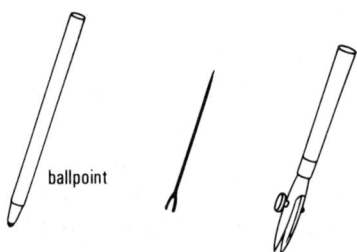

ballpoint

3. *The old fashioned ruling pen* used by draughtsmen rules parallel double slits very well. The separation of the slits is easily adjusted by the knurled knob that moves the two blades closer together or further apart. Make the adjustment by trial, then run the pen along a ruler. If the width of each slit should be greater, grind the tips of the blades a little. *This is probably the best method for the teacher's private reserve of ready-made slits.*

4. *Special ruling device.* Insert the glass slide in the ruling device and rule one slit in the Aquadag with a blunt needle or pin held up against the cross-piece. (The edge of a small screwdriver could be used. A fine biro works well in practised hands.)

To rule a double slit: rule one slit; then displace the slide slightly by turning the thumb-

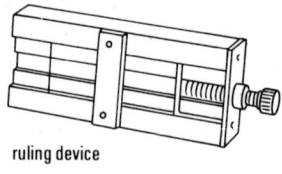

ruling device

---

ingenuity of physicists. There are many good schemes and devices. For our qualitative class experiment now there are two dangers:

(*i*) If the method of ruling strikes pupils as both necessary and *special*, some may even think the phenomenon arises from special ruling.

(*ii*) More generally, if the ruling is complicated, some pupils may miss the wood for the trees and remember ruling rather than seeing fringes.

(An example of both dangers together is the use of the traditional 'bucket-and-cylinder' device for Archimedes' Principle where the logic obscures the real issue for young pupils.)

Therefore we recommend a simple direct method with a ruler and a blunt pen or needle.

We also mention special devices that would help teachers to make good slits quickly for a private stock to help very discouraged pupils.

screw on the end of the device; then rule the second slit.

SIZE AND SEPARATION OF SLITS FOR YOUNG'S FRINGES. In a large lab, slits $\frac{1}{2}$ mm apart, centre to centre, will give fringes that are wide enough for pupils to see easily, *provided the screen is several metres from the double slits.*

If each of the two slits is itself about $\frac{1}{4}$ mm wide, only two or three bright fringes will be visible – because such a wide slit does not spread the light over a broad diffraction pattern. However, those fringes will be brightly illuminated.

If each slit is only about $\frac{1}{8}$ mm wide (and the slits are still $\frac{1}{2}$ mm apart) the pattern of fringes will be just as coarse; but *more fringes will be visible; and the pattern will look much less bright.*

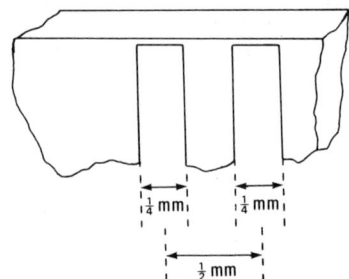

In a small room, it may be necessary to rule the two slits closer together to make the fringes far enough apart to be seen easily.

There is no need to make the rulings conform to any special values. The suggestions above are merely general guides.

PRE-TESTING SLITS. When pupils have ruled several pairs of slits they should bring them for inspection before they use them.

The best test would be to set up the slits as in the actual experiment and look at the fringes; but that would bring the teacher into the pupils' main observation. Instead, the teacher should make a quick test (in which practice will give skill) as follows.

Hold the slits just in front of one eye and look at a distant line-filament lamp. Then the eye acts as a 'telescope' focused more or less for infinity. One sees Young's fringes and can judge them for brightness and separation. *This arrangement should be a teacher's test: it should not be used by pupils,* because the use of the observer's eye in this way is very confusing and likely to spoil the essential message of the experiment.

After some experience in testing slits, teachers

may prefer a simpler test, which pupils will appreciate: judging by eye. Have a small translucent screen at hand, brightly illuminated from behind, and hold the slide with slits in front of it. (One could have a specimen 'recommended' slit already hung on the screen for comparison – though that might not seem good Nuffield teaching.)

HOLDERS FOR DOUBLE SLITS. When a pupil has ruled a 'good' pair of slits, he must clamp his glass slide where it will intercept a beam of light. Ordinary clamps will be awkward for the pupil – who must adjust the slits to make them vertical – so we suggest using a comb holder from the ray-streak kit to hold the slide. If its jaws let the slide slip, line them with plastic tape. The comb holder should itself be held by a clamp on the retort stand.

Alternatively, use the bulldog clip held in a clamp.

SPACING AND SIZE OF SLITS FOR YOUNG'S SLITS: NOTES ON THEORY. The amount of light reaching the fringe pattern is determined by the width of each slit of the pair. The wider each slit, the more light.

But the width of the patch over which the fringe pattern is spread (by diffraction from each slit) varies inversely as the width of the individual slits. Therefore, wide individual slits give more light concentrated into a narrower region, making the fringes much brighter. On the other hand, the narrower that bright patch formed by diffraction, the fewer fringes there are visible – and pupils need to see several dark and bright fringes to be convinced. A compromise is necessary.

The closer the two slits are together centre-to-centre, the greater the spacing from fringe to fringe, and the easier the fringes are for pupils to see.

Therefore, we should aim at using a double slit with the two slits each as wide as possible and with as small a separation between them as possible. Those two conditions would lead to two slits overlapping and merging into a single slit if we pushed them too far! The widest slits allowable for a reasonable picture of fringes seem to be slits of width x whose centres are a distance $2x$ apart. That is, two slits of width x with an opaque region of width x between them. That arrangement will give three bright fringes with two dark fringes between them, and little illumination in regions beyond that.

If the slits made are too far apart or themselves too wide, the central maxima of the diffraction patterns may not overlap, then no fringes will be seen.

Very thin slits are ideal for making a broad display of many fringes: but that display would have to be observed with a magnifying glass, or photographed, because the fringes are both fine and faint – then the cogent simplicity is lost.

INTERFERENCE AND DIFFRACTION

Let pupils proceed immediately to look at interference of light in their own experiment. This first look is for acquaintance, without measurement, before they know what to expect. This should be a memorable experience.

Young's fringes When two streams of light arrive at a screen (from a single source) pupils probably expect to see a brighter patch where the streams overlap. (One lot of bullets + another lot of bullets should make still more bullets.) Now let them find a strange property of light by their own observations in a class experiment.

They let light from a line filament lamp shine through a pair of slits. The two streams that emerge continue to a screen far away, and make bands of light and dark – 'interference'.

At this stage pupils will get so much more from doing the experiment themselves that the trouble of arranging a class experiment is well worth while.

The outcome we hope for is a pupil's boast:
'I have seen light waves interfering: light + light making black as well as bright.' (Better still if he can add, 'and I can take it home and make it work there'.)*

We hope that the only difficult bit of manipulation, ruling the slits, will be done by each pupil but will be made as simple as possible – with help given if necessary. If pupils make the double slits themselves, they know there will be two streams of light.

A strong plea: no lenses There should be no lenses, or the important message will get confused. (Young's fringes are traditionally shown with an eyepiece to observe them. That would spoil the

* The J. Willmer fund will pay for loss or damage. (See footnote, p. 115.)

simple cogency of the experiment for pupils. They know that lenses can introduce peculiarities – particularly if the observer is not sure where the observing eye is being focused. So we plead strongly for a clear experiment without lenses. But that needs a bright source and large distances.)

An important experience We hope teachers will not regard this as a difficult experiment to prepare and run. It gives important qualitative evidence and it prepares for the measurements just ahead.

---

†Class experiment 64
Light from a pair of slits: Young's fringes 'qualitative)

Outline Light from a ray-streaks lamp meets two slits, very close together, in a black screen 1 or 2 metres from the lamp. The light passes through the slits, and continues, spreading a little, to a translucent screen several metres further on. Pupils standing *behind* that screen see interference bands where the two patches of light overlap. There are no lenses, so the wave property of light seems inescapable.

Apparatus

From double slits kit:	item 97
4 dozen microscope slides	97A
1 bottle of Aquadag	97B
1 soft paintbrush $2\frac{1}{2}$ cm wide	
32 short metal rulers (or plain microscope slides)	
32 pins or pens	
1 reel masking tape	105/2
From ray-streak apparatus:	94
2 to 4 lamps, holders, and shields	94A
16 holders for combs, or bulldog clips	94F
2 to 4 transformers	27
16 retort stands, bosses, and clamps	503–506
8 screens of greaseproof paper about 30 cm × 30 cm	
plastic tape (to line bulldog clips)	
2 to 4 home-made 'limiting screens' of cardboard or wood	

Pupils share lamp and translucent screen in groups of 4. (There is enough apparatus for pairs, but more than 8 beams of light traversing the room may make the running of the experiment difficult. On the other hand, there should not be still fewer

light-beams and larger groups: there would be confusion of waiting; and pupils' independence would be lost.)

Each pupil should make, and use, his own pair of slits.

Preparation

Arrangement of the lab. With eight groups, each using lamp and slit-holder and translucent screen, time and trouble can be saved – and confusion avoided – by careful planning.

The actual arrangement must depend on the shape and size of the room and the positions of benches, etc. The suggestions below are offered as hints.

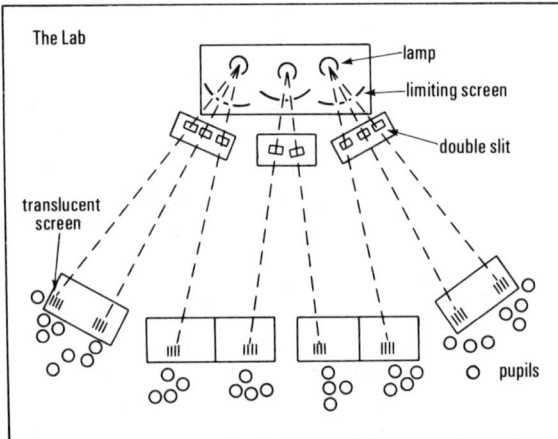

Suppose the lab is about $7\frac{1}{2}$ m by $12\frac{1}{2}$ m. To keep the groups apart from each other, place the lamps near the centre of one long wall. And place the translucent screens for observers all along the opposite wall, continuing round the corners to part of each end.

The translucent screens should be spaced at least a metre apart so that they are separated enough for each to have a quartet of pupils round it.

Preparing the lab. If possible, arrange the lamps beforehand.

If the lamps are high up, the double slits and the translucent screens can be raised to pupils' shoulder level, making adjusting easier and observing more comfortable. That also lessens the danger of table-top reflections. *So it is best to place the lamps (with limiting screens) on stools standing on a bench.*

106

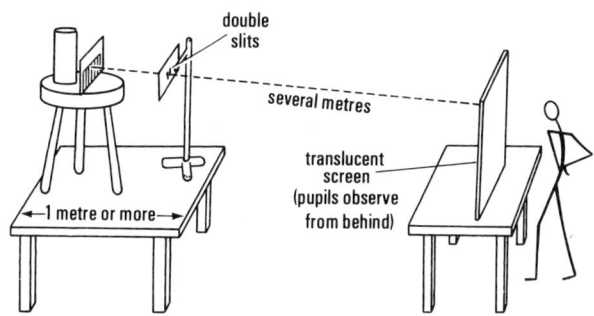

double slits

several metres

translucent screen (pupils observe from behind)

←1 metre or more→

If the translucent screens can be set up beforehand, it will help the running of the experiment. Turn on the lamps and install 'limiting screens' to cut off stray light. Follow the emergent beams of light across the room to translucent screens 4 to 6 m from the lamp. There must be space for four pupils *behind* each screen.

That in turn will define the placing of the double slits. Pupils will clamp their pair of slits on a stand at least 1 m (better, $1\frac{1}{2}$) from the lamp. If those stands are arranged beforehand, set the clamp on each at the right position to intercept the light on its way to the translucent screen.

Make sure that light does not reach the screen by reflection from table-tops on the way.

Preparing slides for slits. These need time to dry, so the coating should be done beforehand. Make one Aquadag-painted slide for each pupil, and 50% more as spares. *Each pupil should rule his own slits.*

A few pupils may become seriously discouraged. For them, after a few attempts, the teacher may decide that an offer of ready-made slits will provide comfort and ensure success. So, when the slides have been painted a few should have 'good' pairs of slits ruled on them – to be kept in reserve.

Procedure

Keep the room lit while each pupil makes several pairs of slits.

Then darken the room fully and help pupils to orient their double slits parallel to the lamp filament.

Pupils follow these instructions:

⋆　　⋆　　⋆　　⋆　　⋆

Use a lamp with a straight filament as the source of light, near one end of the room. Make sure its filament is vertical.

Ask for a small sheet of glass coated with black paint. You can scratch two slits in the paint very close together. Then, if light shines on the black sheet, two lots of light will get through the slits and make two patches of illumination on a screen at the other end of the room. If the slits are thin and close together the light from each slit will spread and the two patches will overlap. What will you see there?

To make the slits, hold a ruler across the glass sheet and scratch the black paint away by dragging a blunt pin or a fine ballpoint pen along the ruler.

To make the second slit, hold the ruler there but tilt the pin or pen and drag it along again. The slits need to be *very* close, about $\frac{1}{2}$ millimetre apart, or even closer.

Make several pairs of slits on one sheet of glass. Then ask your teacher to look at them and choose a pair that will behave well.

Place the slits in a clip, one or two metres from your lamp. Make sure the slits are vertical, parallel to the lamp filament.

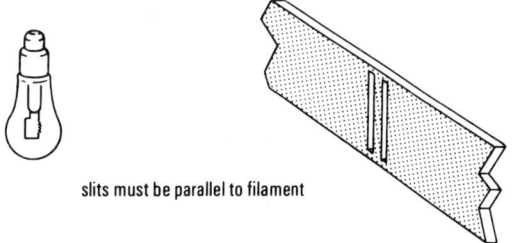

slits must be parallel to filament

When the room is dark, hold a piece of paper just beyond your pair of slits and see the light that comes streaming through and spreading out a little.

Then go as far away as possible – it must be several metres – and hold a translucent screen of greaseproof paper there. Place the paper to catch the light that comes through the pair of slits.

Go round *BEHIND* the screen and look at the bright patch where the two lots of light overlap.

What do you see? What do you think that tells you about light?

⋆　　⋆　　⋆　　⋆　　⋆

Each pupil looks at the pattern with his naked eyes. Some will need a reminder: 'Stand some distance back, as you would in reading a book.'

Young's fringes do two things for us: they provide evidence of the wave behaviour of light, and they yield an estimate of wavelength. The first is by far the more important in our present teaching.

Tell pupils that the pattern of stripes (fringes) is somehow made by the two patches of light arriving together. It is called Young's fringes after Thomas Young, who claimed it as a test for light waves.

Discussion After pupils have made Young's fringes themselves, hold a general discussion. One might say:

'Extraordinary. Bands of black and white, like a zebra. If light is a stream of bullets, could you have:

LIGHT + LIGHT, making LIGHT, in some places; and LIGHT + LIGHT, making darkness, in other places?

You could have:
BULLETS + BULLETS, making MORE BULLETS;
but could you have . . . ?

Have you seen anything in physics where two lots of something arrived and made a *big* result in some places and made *nothing* (cancelled out) in other places?'
Ask whether waves could make such patterns. Suggest that pupils should look at water ripples from two sources, where the ripples overlap.

Pupils get out ripple tanks and look at the wave pattern from two small sources vibrating in phase. They may need some help in seeing the patterns that lead to Young's fringes.

The strange name, 'interference' At whatever stage the name *interference* enters the discussion, we need to say it is an unfortunate choice: waves do *not* upset each other; their effects simply add. The terms *constructive interference* and *destructive interference* are slightly less confusing but the word itself remains misleading.

†**Class Experiment 65**
Ripples from a pair of sources: 'Young's fringes'

Apparatus

8 ripple tanks	item	90
8 lamps		47
8 transformers		27
8 vibrating bars		90L
16 point dippers		90G
32 hand stroboscopes		105/1

Pupils work in groups of four, but each has a stroboscope.

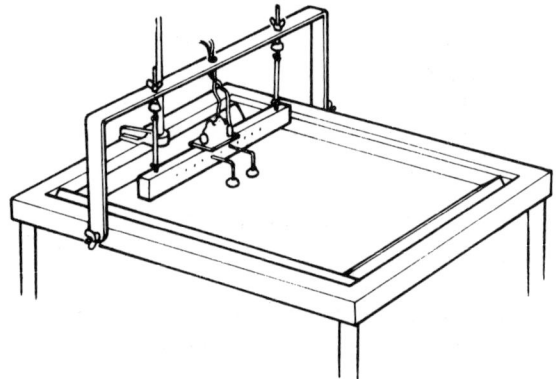

Procedure

Pupils set up the ripple tanks as usual (for general instructions see *Teachers' Guide Year 3*, page 10).
Pupils follow these instructions:

 ★ ★ ★ ★ ★

Set up your ripple tank, with the vibrating bar just above the water.

Install two dippers about 3 cm apart on the vibrating bar. Run the motor as slowly as possible (about 10 revs/second). Look at the pattern in the shadow of the tank. Can you see some curved lines where there seems to be no motion, and other lines where there is much motion? All those lines are due to two lots of ripples adding their effects.

Make the motor run faster and use a hand stroboscope to make the pattern visible all the way out across the tank.

If you like, try changing the distance between the dippers.

 ★ ★ ★ ★ ★

Diffraction of ripples While ripple tanks are in use, pupils should also look at 'ripples passing through gaps' because we shall soon proceed to the diffraction of light – not only important as evidence for waves but also needed in dealing with Young's fringes and (in Chapter 9) gratings.

---

†**Class Experiment 66**
Straight ripples pass through a gap: diffraction

Apparatus

8 ripple tanks	item 90
8 lamps	47
8 transformers	27
8 vibrating bars	90L
16 straight barriers	90D
16 side barriers (blocks of wood)	
32 hand stroboscopes	105/1

Waves coming round the outer ends of the barrier are troublesome. They must be blocked off with side barriers. Avoid high frequencies which may start the barriers themselves vibrating.

Pupils work in groups of four but each has a stroboscope.

Procedure

Ask pupils to try letting straight waves go through a narrow gap. Pupils follow these instructions.

     ★   ★   ★   ★   ★

Adjust the vibrating bar so that it is just in the water. Then it will make straight waves.

Build a wall of barriers about 5 cm from the bar. Leave a 10 cm gap for waves to pass through. Run the motor slowly and watch the waves that go through.

Run the motor fast and look at the pattern with a stroboscope.

Then narrow the gap to about 1 cm. What do the waves beyond the barrier look like now?

Try a still narrower gap.

     ★   ★   ★   ★   ★

Geometrical demonstration in teaching interference The idea of waves adding when they are in phase to give a large wave, and 'interfering destructively' when they are in opposite phase, is easy for adult physicists but strange to many pupils. It is easy for pupils too, once they have grasped the essential idea: but that seems to need some extra demonstration of the difference of phase that arises from different paths.

The demonstration with plastic wave strips does that.

Demonstration 67
Plastic wave model for interference

Apparatus

2 wave strips	item 126
2 retort stands	503–504
4 bosses	505
2 small G-clamps	44/1

Preparation

Support the wave strips with two retort stands. The best way is to secure a block of wood or plastic to the side of each strip, at one end. Drill a vertical hole three-quarters of the way through each block, to receive the top of a short retort stand.

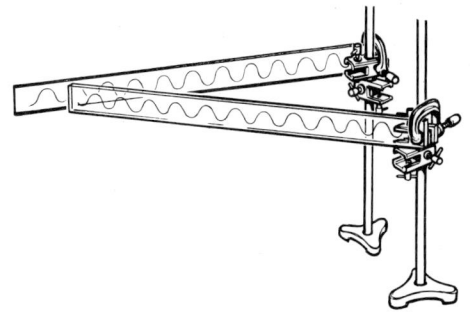

Each strip extends out horizontally, but its face with the pattern is vertical. Place the two stands 3 or 4 wavelengths apart.

Procedure

Hold the free end of each strip and make them cross. Swing them to show what happens at various cross-over points.

A short description with words and hands may help too.

'This wave arrives and makes things go

FLIP–FLAP, FLIP–FLAP . . .

then the two waves added together make things go

FLIP–FLAP, FLIP–FLAP . . .

But when we move to another place where one wave has travelled half a wavelength further than the other, they arrive out of step. The first wave makes things go

FLIP–FLAP, FLIP–FLAP . . .

and the other wave makes things go

FLAP–FLIP, FLAP–FLIP . . .

and then the total of the two is . . .?'

Simple model of wave addition Instead of the 'waves' engraved on strips, pupils use thin strips cut from a piece of corrugated cardboard. This works better if pupils have already been shown by a demonstration what they are going to do. Then, as well as using it in class, pupils might take it home and explain interference patterns – not very amusing alone, but a magnificent chance for pride if the real Young's fringes experiment can be borrowed as well.

Class Experiment 67X
Model of wave interference with corrugated cardboard

Apparatus

16 drawing boards   item 551
16 ordinary pins
32 drawing pins
corrugated cardboard

Preparation

Cut the corrugated cardboard in narrow strips, each about 20 cm long by $\frac{1}{4}$ cm wide, two strips per pair of pupils. Have spares available for taking home.

Procedure

Cut two long narrow strips from the cardboard. Place them *on their sides* on a drawing board. Pin

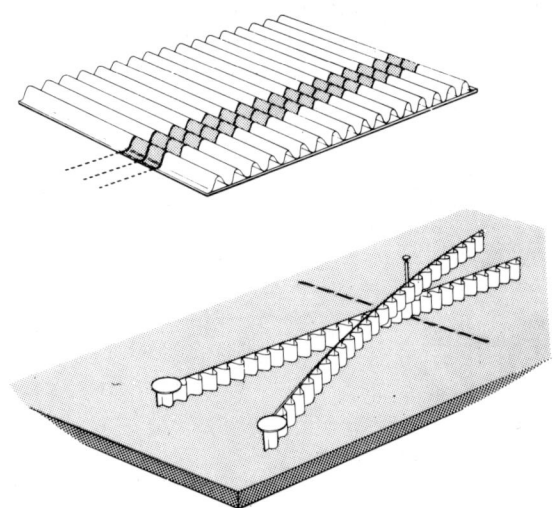

each strip to the board by a drawing pin through a wave-hump near one end.

These anchored ends should be a few centimetres apart. They represent two sources of waves, like the two slits for Young's fringes.

Near the other end of the board mark a line to represent the screen where you expect to find a pattern of 'bright' and 'dark'.

Pull the two strips taught and stick a single pin through two wave-humps (one hump of each strip) to find a place on the 'screen' where the waves add up to 'bright'. Find several such places on the 'screen'.

This is an experiment pupils could show at home to explain Young's fringes if they show the real experiment with light.

DIFFRACTION OF LIGHT: A QUICK LOOK

Pupils should certainly see examples of diffraction – not just pictures in a book but real objects and their 'shadows' – not only for arguments about light but also for wonder. They probably expect to see sharp shadows cast by light from a point source, in *all* circumstances. '*Light travels in straight lines*' is learnt only too thoroughly, with never a thought about the philosophical assumption in 'travels', still less any doubt about 'straight lines'. And talk of 'bullets' as a model for light will reinforce that. So, although diffraction should not bulk large and take much time from the interference experiment that follows, pupils should take a quick look, for surprise.

Provide a selection of objects to cast shadows. Pupils should first look at the objects close by to see the things that are to cast the shadows. Then on a screen far beyond they see the 'shadows'.

The wavelength of visible light is so small that the screen must be placed far away, the objects must be fairly small, and the source must be *very* small, a tiny pinhole. Even with a very bright filament behind the pinhole, the illumination on the distant screen is very faint and pupils viewing an unexpected pattern in a dark laboratory need special help. This is provided by a *translucent screen which they view from behind.*

The traditional alternative to the screen is an eyepiece or other arrangement of lenses – often allowing viewing at much smaller distances. We urge teachers strongly to avoid all lenses – except the eye itself – because it is all too easy for beginners to blame the lenses, letting confusion and mystery take the place of wonder and surprise.

The room must be fairly dark. Yet if we try to show diffraction in a completely blacked-out room, we may defeat our own aim: pupils cannot see what is happening and there may be difficulties with discipline.

In these experiments, where pupils need time for their eyes to adjust, remember that darkening the room usually cuts off fresh air as well as light; and as the atmosphere grows warmer and more moist, pupils feel less comfortable and discipline is likely to suffer – for a simple physiological reason. So we do not recommend complete blackout, only half or three-quarters darkening of the room.

Class Viewing 68
Sharp shadows: Diffraction

Apparatus

1 compact light source	item 21
1 metal plate with pinhole (1 mm)	
1 large translucent screen of greaseproof paper†	
3 retort stands and bosses and clamps	503–6
1 piece of plate glass (15 cm × 15 cm) to hold objects	
objects‡	
hard wax to attach objects to plate	
1 variable voltage supply	99

† See the note on p. 103 for Young's slits equipment.
‡ Suggestions: sewing needle, pin, metal plate with holes of diameters ranging from 2 or 3 mm to 1 cm; a human hair (very popular); a metal ball 3 mm diameter (5 mm at most). For an honest diffraction picture from the last two, a small pinhole source is essential.

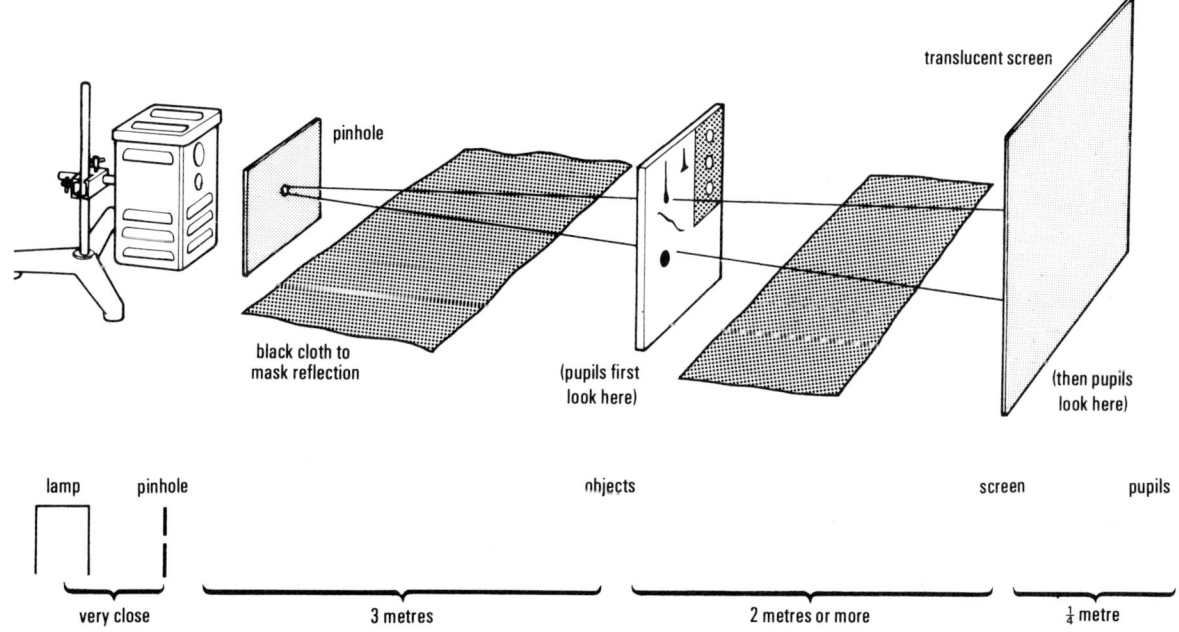

111

Preparation

Place the compact light source at one end of the lab, with the pinhole just in front. Place the translucent screen near the other end of the room.

In the middle between them place the objects to cast shadows. The objects should be far from the lamp, at least 3 m away. The screen should be at least 2 m beyond the objects.

Stick the objects on the glass plate. Place the plate so that pupils can easily go to it, look at it closely, and hold a sheet of paper just beyond it, to see the objects.

Shield the screen from light reflected by table tops.

Procedure

Make the room three-quarters dark.

(*i*) *The objects* Pupils stand near the collection of objects and look at them. They see sharp shadows on a piece of paper just beyond.

(*ii*) *The diffraction effects* Then pupils move to the translucent screen, *go round behind it*, and look at the shadows there. Remind them to hold their heads $\frac{1}{4}$ metre or more behind the screen – 'as in reading a book'.

The shadow of a disk. To a physicist, the strangest shadow of all is that of a small ball or disk; there is a white spot in the centre of the shadow.* One can just see it, in a long, dark room, if one expects it: but our source is too large and pupils will probably miss it unless the source is small.

(*iii*) *Diffraction by a slit* Change to a set of three prepared slits, wide, medium, and very narrow. Suitable widths are 3 mm, 0·3 mm and 0·03 mm. The narrowest needs no microscope to check its width; judge it by its diffraction spread which should be a few centimetres on a screen at 2 m. The slits could be ruled on a coated glass slide, the smallest with a razor blade.

(Avoid a V-shaped slit or a variable slit. Those belong in A Level and would be confusing now.)

---

* When Fresnel as a young man submitted his paper on the wave theory of light to the Academy of Sciences in Paris, Poisson read it and objected at first that it must be wrong because he saw that it would predict that white spot. Fortunately the spot was observed, and later photographed. In an earlier age it might have led to an accusation of witchcraft.

Evidence Now discuss with pupils the qualitative evidence for light being a wave motion. There is already a quantitative requirement – as Newton knew. This requires the wavelength to be very small, and the frequency very great, a million times greater than for VHF radio.

At the same time, leave a warning question unanswered: '*Are you now sure that light always behaves like waves and never like bullets?*' Pupils will soon meet the photoelectric effect and then extend a view of dual behaviour to electrons and all material objects.

{**Explaining diffraction?** Other patterns made by light passing various objects can be predicted or 'explained' as wave effects. Neither the black spot in the centre of the bright patch made by light through a hole at some distances, nor the white spot at the centre of the shadow of a disk at all distances, makes much sense if we treat light as a stream of bullets. Nor do these make sense easily on a wave basis without considerable geometrical discussion; so, when pupils see these we can only assure them that they are in fact the effects that we should expect of waves.}

{Attempts to discuss Fresnel zones, etc., lead to confusion at this stage and are, in fact, rather questionable optical teaching at later stages. All the more reason for giving great importance to Young's fringes which pupils can see with light, and compare with water ripples, and understand fully geometrically.}

YOUNG'S FRINGES: ESTIMATE OF WAVELENGTH

The diffraction exhibit can provide a time for digestion of the first experience of Young's fringes. After that, ask pupils to make a measurement to estimate the tiny wavelength of light. For that estimate, pupils must go through some geometry. A fast group should make the geometry their own, first being shown it and then learning it so well that they could teach it to others.

An average group should be shown the geometry – for the good reputation of science – so that they can say, 'We have seen that and are satisfied that it is sensible' – and then they may use the result.

The geometry is essentially similar to that for a diffraction grating. The latter is a Young's-fringes arrangement with many slits, making its fringe pattern at infinity or the equivalent. Teachers familiar with the use of plane diffraction gratings may feel that the grating geometry is simpler; and they may plead for a grating instead of Young's double slit. But here, where we want to get light fringes experimentally, directly, and convincingly, we urge teachers to deal with Young's fringes first.

To make the geometry simple, and to prepare for the grating, draw a more realistic diagram than the usual one. Show the two slits a small distance apart and place the screen an enormous distance away. For example, sketch II, page 114 shows a pair of slits $\frac{1}{2}$ mm apart, magnified 30 times. Then the screen with Young's fringes, actually 2 m away, must be imagined 100 m away. Pupils can see that the wave paths from the two slits to a place on the screen are almost parallel.

Sketch II shows the tiny path difference. If that path difference is one wavelength, L, the two contributions arrive at the screen in phase and make a bright fringe there.

Class Experiment 69
Young's fringes: estimate of wavelength

Although this uses the same equipment as the qualitative experiment, it will go better with a fresh start – so we suggest it should come after an interlude of diffraction.

Apparatus

The same equipment and arrangement as for **Class Experiment 64** with the following additions:

16 centimetre rules	item 502B
16 metre rules	501
transparent $\frac{1}{2}$ mm scale	7E
or microscope and mm scale to measure distance between slits	

Preparation

The same as for **Class Experiment 64.**

Procedure

Pupils follow these instructions.

*   *   *   *   *

Obtain a good clear set of Young's fringes as you did before. (If you kept a good pair of slits from your previous experiment, use them again. Otherwise make several new pairs and choose the best – you will find that is a much quicker job in this second attempt.)

Measure the spacing of the fringes on your translucent screen. Go round behind the screen with a scrap of paper. Hold the paper against the screen and make pencil marks on it at each bright fringe. Then carry the paper out into daylight and measure the distance between extreme marks. Divide to find the distance from one fringe to the next.

If you like, compare your estimate with the estimates of your partners and take an average.

Measure the distance from your pair of slits to your translucent screen. Since your fringe measurement is likely to be fairly rough, it would be unscientific to measure the large distance very accurately – measuring to the nearest few centimetres is good enough.

Measure the distance between your two slits, centre to centre. This is the most difficult measurement of all. But remember that you are going to arrive at something far smaller still, the wavelength of light. So any rough measurement is still very valuable.

You may be able to compare your double slit with a millimetre scale under a magnifying glass or even under a microscope. Or you may be able to hold it in a slide projector beside a transparent scale. Measure it *somehow* even if you are partly measuring, partly guessing.

The calculation Look at sketch I. This shows the arrangement but it is greatly distorted: it makes the distance between the slits much too large compared with their distance from the screen. For the central bright fringe at A, the wave contributions from the slits S_1 and S_2 arrive in step. The paths are equal. So $S_1A = S_2A$.

For the next bright fringe at B the wave contributions again arrive in step, but one path is a whole wavelength greater than the other. $S_2B = S_1B + L$ or $S_2B - S_1B = L$, one wavelength.

Now look at sketch II which is greatly magnified but not distorted. Imagine the two wave paths continuing out to the screen and meeting

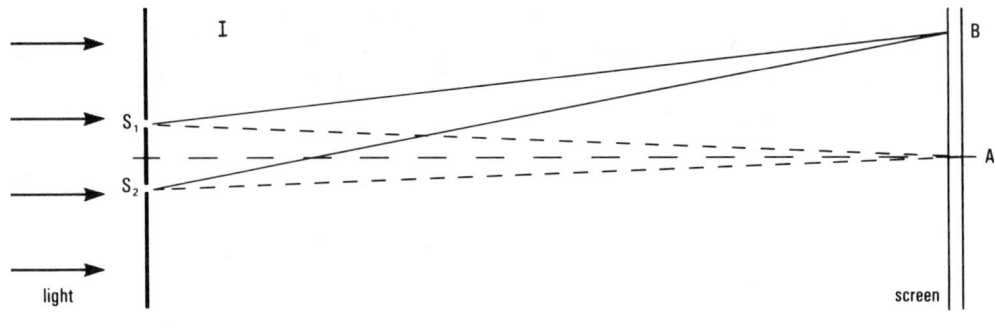

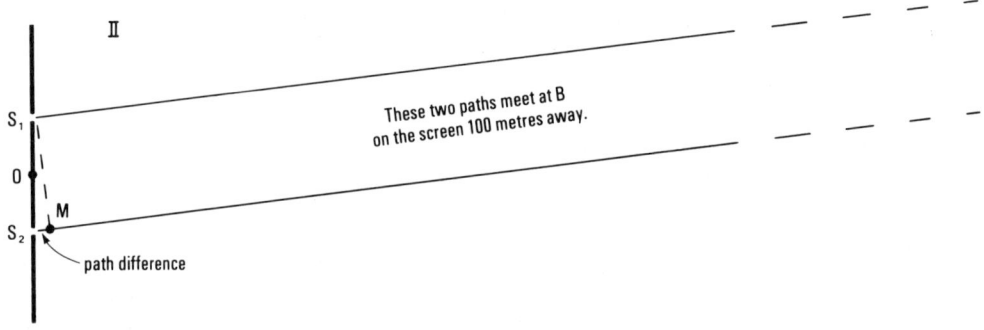

These two paths meet at B
on the screen 100 metres away.

path difference

there at B. (For a $\frac{1}{2}$ mm distance between slits magnified to $1\frac{1}{2}$ cm in sketch II, the screen would have to be shown 100 metres or more away! Can you see from that how very near to parallel the wave paths must be?)

On sketch II, S_1M cuts off the extra path from S_2. We have drawn S_1M practically perpendicular to both wave paths since there is so little to choose between their directions.

Then S_2M is the extra wave path for the light from S_2. For the first bright fringe out from the central one, that extra path must be one wavelength.

$$\therefore S_2M = L$$

Sketch III attempts the impossible, to maintain the magnification and yet show the screen on the paper! In sketch III, the triangles S_1MS_2 and OAB are the same shape ('similar') because OB from the mid-point between the slits is drawn accurately perpendicular to S_1M, and therefore the angle S_2MS_1 is *almost* $90°$. In the large triangle OAB the angle OAB is of course *exactly* $90°$.

Use y for the fringe spacing, AB.
Use d for the distance between the slits.
Use D for the distance OB from slits to screen.
Then, in those two triangles:

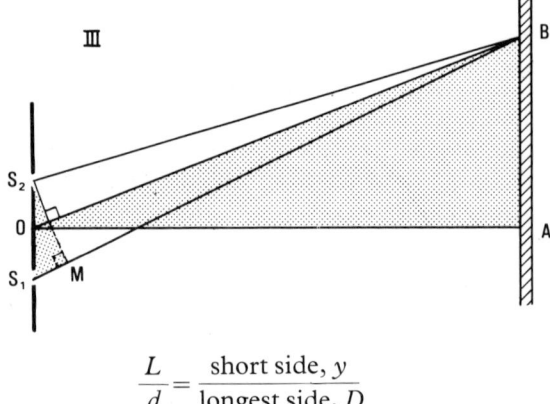

$$\frac{L}{d} = \frac{\text{short side, } y}{\text{longest side, } D}$$

$$\therefore \text{wavelength } L = \frac{y}{D} \times d$$

Draw your own sketch and label your measurements on it. Then calculate the wavelength of light.

Your result will be an average for the many colours of white light, but since there is plenty of green light in it, and green is the colour to which your eyes are most sensitive, you might call it the value for green light. (If you prefer, start again, using a green filter; then indeed you will be measuring the wavelength of green light.)

This is not a precise measurement. It is a case of 'desperate measures in desperate circumstances', where even a rough estimate is very valuable because it tells you which county you are in. It tells you how very small the wavelength of light is.

★   ★   ★   ★   ★

The point of making this estimate is *not* precise measurement. We are moving into the microphysical world where a rough estimate is very valuable. And this is doubly valuable if the pupil makes the estimate himself from his own measurements.

On this view, any result between 200×10^{-9} and 1000×10^{-9} m is still in the right order of magnitude and well worth having. We believe it will be more valuable at this stage than a more precise measurement done as a demonstration.

In fact, the wavelength of light is something to take home as a precious discovery. We hope that pupils will literally take it home with pride, and we trust that schools will encourage them to take home a pair of double slits (or the materials for making them) and a lamp, to show people they can measure the wavelength of light themselves. For home use, where complete darkness can be arranged, perhaps in a long corridor or a cellar, an ordinary 12-V lamp will suffice (preferably run at extra voltage to make it still brighter). We hope that schools will lend the necessary transformer for such a lamp.★

THIN FILMS

As a final look at light waves interfering, pupils should see the effects of light reflected from the two surfaces of a thin film.

A soap bubble shows irregular patterns,

because the thickness is irregular; but its colours remind us – as Young's fringes should have shown – that wavelengths are different for different colours.

Demonstration 70
Soap films

Apparatus

1 beaker, 400 cm^3	item 512/2
a large beaker, 1000 cm^3	513

1 copper wire frame
fresh soap solution (from toyshops) or detergent
glycerine
piece of carpet made of synthetic fibre
lamp and screen

Preparation

For **a**, bend a frame of copper wire, 16 or 18 SWG, as shown. The vertical circle at the top should have a diameter at least 5 cm, preferably 7 cm or more.

Put soap solution in the 400 cm^3 beaker. Soap-bubble liquid from toyshops does very well. Or make a mixture of Stergene and water. (A dilution of Stergene of 1 in 10 gives a rather streaky pattern but the film is strong. 1 in 1000 gives a film with closely spaced fringes but the film is weak. A dilution of 1 in 100 is probably best. Add glycerine to make the film stronger – but the colours will be poorer.)

Procedure

a. Dip the frame in soap solution to make a film. Use this film as a mirror to reflect light from the sky to pupils. As the film drains it thins, and interference bands appear.

★ This linking between school physics and other people at home can do so much for the development of an understanding of science – and thence for the promotion of good teaching – that we plead strongly for this kind of home experiment. But now we should let it change from simple practical things, like making crystals, to a glimpse of a great experiment that is bound up with theory.

While we consider home experiments very important, we recognize the difficulties that school authorities may encounter over lending out apparatus which they feel may get lost or broken. Since we regard home experiments as so important, we offer at present to underwrite such home lending through a special fund. If teachers lend lenses, lamps, transformers, etc.,

and find that they cannot get them back, or the apparatus comes back damaged or broken, they should apply to:

The J. Willmer Home Experiments Endowment, Association for Science Education, College Lane, Hatfield, Hertfordshire, AL10 9AA.

The General Secretary administering this fund will only ask whether the apparatus went on loan with permission, whether the class is following a complete Year of our Nuffield O Level Physics programme, what was damaged, and the cost to be met. He will not want to know the name of the pupil and he will not want the usual formal details of a report of damage. The cost will be reimbursed happily.

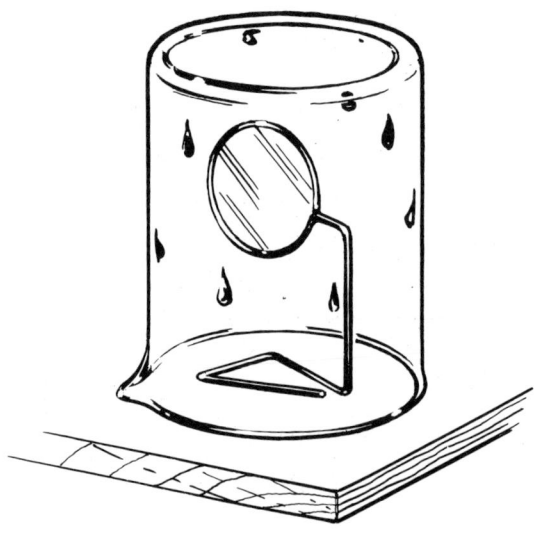

The film will last a long time, even when thinned, if evaporation is discouraged. Place a large beaker, wet inside, over the frame carrying the film as it stands on the bench-top.

b. Blow a soap bubble and catch it on a small piece of carpet made of synthetic fibre. Place a big beaker over the resting bubble. Arrange a bright white screen behind the beaker (or a lamp and large white lampshade above).

Just before the film breaks, the thinnest region becomes invisible. That 'black spot' is spectacular.

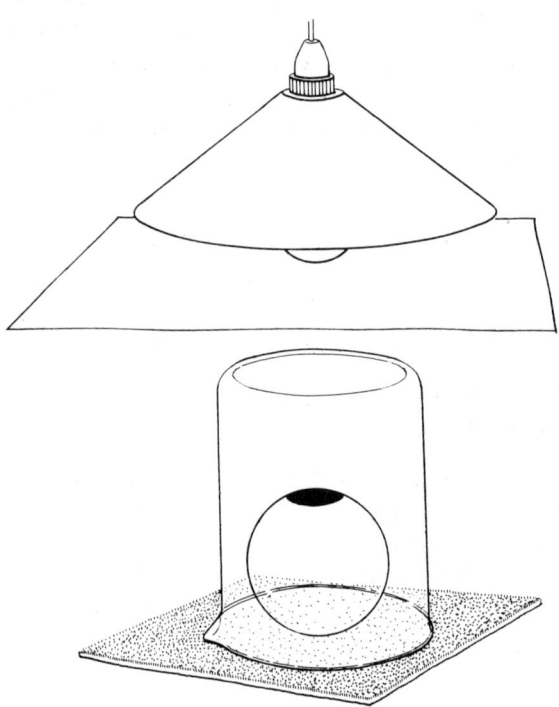

A slight draught makes the experiment more convincing and dispels any idea that the black region is a hole in the film. The most convincing test is to poke the black region with a piece of chalk – a good way of breaking any soap film.

Pupils would expect the light-waves from two surfaces very close together to reinforce, not annul. Here we must say frankly to pupils, 'There is a good reason for the black spot but it is too difficult to explain'.

Avoid saying, 'The film is too thin for light waves to show it' – a thin film of absorbing material *would* be visible.

Air wedge Pupils should look at an air film sandwiched between two plates and illuminated by sodium light (or by a mercury arc with a green filter). This will see fringes caused by 'interference' of light reflected from the two *inner* faces of the glass sheets – the outer faces of the sheets are too far apart to show any noticeable consistent pattern by their reflections.

Class Experiment 71
A thin film of air - interference with air wedge

(This simple experiment should be done by each pupil holding a pair of plates individually. A demonstration for pupils just to look at would not be worth while at this stage.)

Apparatus

2 Bunsen burners	item 508
16 pairs of plate-glass plates	129
32 bulldog clips to hold glass plates	564
1 translucent screen	46/1
1 lamp (behind screen)	46/2
sodium bicarbonate or common salt	
iron wire to hold salt in flame	
16 pieces of red and green colour filter	205
scraps of tissue paper	

Although a sodium lamp is the easiest source of monochromatic light for this experiment, a Bunsen burner fed with common salt or, better, sodium bicarbonate, on an iron wire, provides a good simple source.

Preparation

Clean the pieces of plate glass carefully beforehand. To test them for reasonable flatness, press them together and examine the fringes by monochromatic light.

An air wedge is formed by spacing the plates apart at one end with a scrap of tissue paper.

Arrange the Bunsens behind a translucent screen (or in front of white screens) to make an extended source.

Procedure

Pupils follow these instructions.

⋆   ⋆   ⋆   ⋆   ⋆

Make a sandwich of two plates of glass with a very thin layer of air between them. That layer of air can take the place of a soap film.

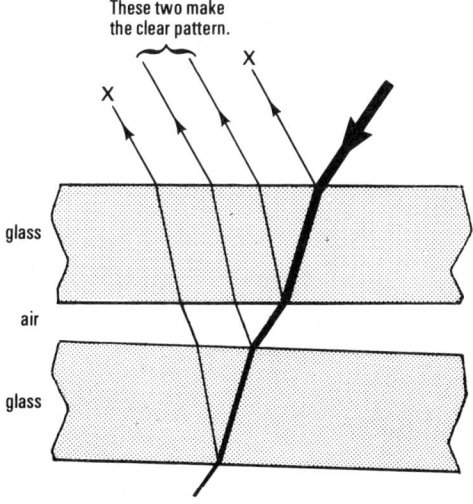

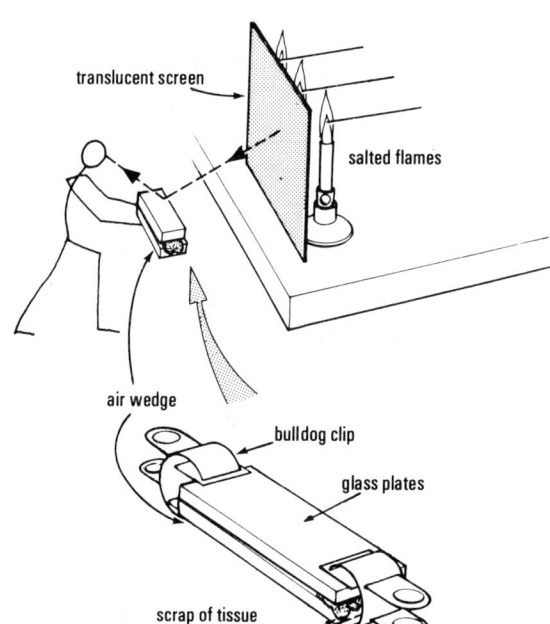

Make the layer of air wedge-shaped by propping the plates apart at one end with a scrap of thin tissue. Clamp the plates together with a bulldog clip at each end, so that the thickness of the air layer is zero at one end and one tissue thickness at the other end.

Hold the sandwich in front of a wide source of pure yellow light; and look at the source's image reflected by the sandwich.

There are four streams of reflected light: two from the inner faces of the sandwich, where glass meets the thin wedge of air; and two from the outer surfaces of the glass plates. The two streams from the inner surfaces have a small path difference (about twice the thickness of the air wedge at each place); and you will see interference bands of bright yellow and black.

(The streams reflected from the outer surfaces of the glass plates have too great a path difference to show an interference pattern noticeably.)

Suppose you counted the stripes all the way from one end of the sandwich to the other. If you knew the wavelength of yellow light (about 600×10^{-9} m), what could you then estimate?

In fact, counting interference fringes like these is the basis of the modern way of specifying the international standard metre.

⋆   ⋆   ⋆   ⋆   ⋆

Home Experiment H72
Air wedge

If pupils live near a street with sodium lamps, they might try looking at a sandwich made with any small plates of glass.

The fringes may be far from straight, but they will be visible by reflection if the plates are held firmly together with thumbs and fingers.

The standard metre Tell pupils that counting fringes like these is the basis of the modern way of specifying the standard metre. Here counting fringes would only tell us the thickness of the scrap of tissue; but the method can be extended.

Note: coherent sources

In describing arrangements to show interference of light we assume that the two sources (the light waves emerging from the two slits), are 'coherent', that they start out in phase. In the early studies of interference, that was a very important condition: ingenious schemes had to be used to derive both streams of light from the same original source. And in teaching one emphasized that and pointed out that two electric light filaments, strung close together, would never show Young's fringes because the sources of light in the two filaments would oscillate quite independently with arbitrary changes of phase. However, that should not be emphasized here – probably not even mentioned unless a pupil asks – because we are discussing a basic demonstration of wave behaviour rather than examining the more advanced details.

Furthermore, we now know how to make atoms in some light sources gang together and co-operate to produce all their light waves in the same phase. Then we have tremendously strong sources of coherent light. These devices, lasers, can pour light through two slits and form very bright, sharp fringes on a screen. The small lasers now available for teaching are safe if the instructions are followed, but we do *not* suggest acquiring a laser for the present experiments because the essence of our aim is to show the wave nature of light *very simply* and lasers look complicated even if they are not.

CHAPTER 9
Diffraction gratings:
Spectra
Electromagnetic spectrum

In this chapter pupils look at gratings and spectra; measure a wavelength, and see a brief sketch of electromagnetic waves.

DIFFRACTION GRATINGS: GENERAL COMMENTARY

{To us as physicists, the diffraction grating is an instrument of great importance. Ruled and crystal gratings have enabled us to show that light, infra-red radiation, electrons, neutrons, etc., all have wave properties; and gratings have enabled us to measure the wavelengths with high precision.}

{With fine gratings we measure the wavelengths of light in line spectra from atoms. The results of such measurements lent support to the quantum theory in its early growth – they provided data and tests for the Bohr atom model. That model and its successors revealed a wealth of information from spectra concerning energy levels and atomic structure.}

{In the analysis of materials, spectroscopy has long been of great value. And it is an essential tool in astrophysics.}

{All these advances in knowledge have come essentially from measurements of wavelengths. We wish we could treat diffraction gratings fully and do justice to those great developments. But, at this O-level stage, we cannot.}

{Take, for example, the passage from line spectra to energy levels in atoms. We would have to build up the problem; show our pupils the background of knowledge and the need for new answers; describe early ideas of atomic structure; show the decoding of spectra series into term formulae; the lead up to the need for a quantum restriction; and at last come to the connecting together and fruitful conclusion. To pile that on the basic geometry of diffraction grating measurement – itself difficult for some – would make too heavy and too long a task now.}

{On the other hand, we must teach the qualitative use of gratings to demonstrate waves because we shall talk of wave behaviour extending throughout the whole of Nature. We shall talk about the wave properties of light; mention the wave nature of X-rays; and describe 'matter waves', thus linking optical grating spectra with electron wave phenomena.}

{Therefore, pupils should do some simple experiments with gratings and try making a measurement. But we should *not* extend such experiments into a series of measurements of line spectra, since we shall not put them to use. The diffraction grating is a topic to treat quickly, encouraging a feeling of success by giving help when it is needed.}

Pupils should do class experiments with a coarse grating in order to see many spectra; then with finer gratings to measure the wavelength of visible light. In most experiments they can view spectra with the grating held close, the eye taking the place of a telescope. Beginners will need some explanation of that.

VIEWING SPECTRA DIRECTLY

{In the usual diagram, plane waves (or a parallel beam of rays) fall on a grating and give rise to parallel beams in various directions thus forming the spectra. We imagine a screen for the spectra at infinity; or we insert a lens to form the spectra on a screen in its principal focal plane.}

{When a pupil holds a grating close to his eye, he is using his eye as the lens, and his retina as the screen. In other words, his eye sorts out various parallel beams of light (or *groups of contributions* to plane waves) and brings each beam to a focus on his retina. Although to physicists this seems optically similar to an open demonstration with a lens, it will seem puzzling to many pupils, even though delightful to look at.}

When pupils have used gratings themselves, offer a quick demonstration in which grating

119

spectra of white light are projected on a wall far away. In an advanced laboratory one adjusts a spectrometer to make, and then analyse, parallel beams of light – plane waves from each point on the slit – so that precise measurements can be made. But such an arrangement of two lenses makes too small a spectrum for a good demonstration. In our present demonstrations we do not aim at precision, so we do not mind if different rays of light from a point on the source meet the grating or prism at slightly different angles. Therefore we use one lens and place it to form a real image of the source at a great distance. We place the grating or prism just beyond that lens.

Looking at various spectra with a grating
Pupils return to their own experiment with a piece of fine grating held close to the eye. After looking at a white-hot filament, they should observe a neon tube (and, if possible, a hydrogen tube arranged to give the atomic spectrum lines), then a slit with a bright sodium flame behind it, and then again a white-hot filament, this time with a red or green filter held in front of it.

GRATING: EXPLANATION OF ACTION

{The wave story of optical images – in contrast with the ray story, which is incomplete and only roughly true – is that all contributions from an object point must reach the image in *the same phase*. It follows that, where the wavelets from all the rulings of a grating arrive *in phase* there will be a bright 'image' of the source. We do not have to claim that these separate wavelets coagulate to form perfect plane wavefronts in several different directions for the pattern of several spectral orders as in the familiar treatment of the action of a grating.}

{We may show that near a 'grating' in a ripple tank we see round ripples emerging but that, further away, we *do* see those portions of the ripples that touch the tangent line travelling on in agreement, making something like a plane wave; and that in other directions the round ripples *seem* to offer such a variety of contributions that they cancel out to practically nothing. This is the same as the traditional use of Huygens' Principle, but it does not start by saying that Huygens is right or continue by wrangling about Huygens's full geometrical story – which students find it hard to swallow.}

If pupils agree that the grating produces something like plane waves in various directions for the different orders of spectra, they can study the simple geometry.

Draw 'rays' from successive slits to a diffracted plane wave front and point out that the extra path when we change from one slit to the next is one wavelength (or 2, 3, ... wavelengths for higher orders). Then the geometry is clear:

$$\frac{\text{one wavelength}, L}{\text{slit separation}, d} = \sin A$$

where A is the angle of deviation of the spectrum from the original beam.

In all this, we have spoken of the grating as made of slits. In a real grating, each ruling is a furrow of complex profile, but it is easier for beginners to think of a grating as an assemblage of narrow parallel slits, regularly spaced a small distance, d, apart, centre to centre, each slit so narrow that light waves spread out widely from it by diffraction and thus contribute to several orders of spectra.

Manufacture of gratings It may help if one gives a description of ruling a grating on glass: a diamond point ploughs a furrow, spoiling the surface of a flat sheet of glass. It ploughs furrow after furrow, usually leaving a small strip of unspoiled glass between adjacent furrows – then pupils can picture light pouring through the unspoiled strips.

In fact, of course, the furrows also let light through, in various phases that compound to some definite contribution – the same from each successive ruling.

Modern *reflection* gratings are ruled on soft metal, the diamond ploughing up the metal to make slanting ridges which provide preferential reflection in some particular direction. A grating that is 'blazed' like that throws most light into the spectrum nearest to the preferred direction.

Cheap gratings for teaching use are made by casting a film of plastic on a ruled grating – perhaps a reflection one – and then peeling it off. If the casting and peeling are done carefully, the replica has the same grating space, d, as the original. The value of d is obtained from the maker, who counted the number of furrows as he ruled them. Therefore, in telling pupils the grating space for use in their measurements of wavelength, we are not arguing in a circle.

THE GRATING EXPERIMENTS

Let pupils look at a bright line filament of a lamp through a piece of fabric, a coarse ruled grating, and a fine ruled grating.

Class Experiment 73
A quick look: fine cloth as a grating

Apparatus

1 compact light source†	item 21
1 L.T. variable voltage supply	59
or 1 transformer	27
1 retort stand, boss, and clamp	503–506
many pieces of woven cloth‡	

† The compact light source should be used, rather than a line filament lamp, since this is a two-dimensional grating.

‡ The cloth should be of a simple square weave with holes between the threads to let the light through. The samples should be large enough for pupils to hold with two hands and try stretching and shearing. (*Pupils' Text* refers to an open umbrella as an encouragement to a home experiment with street lights.)

Procedure

Each pupil looks towards the lamp and holds a piece of cloth in front of one eye.

(Looking at sodium street lamps through domestic net curtains is very effective, especially if there are lamps at various distances.)

Class Experiment 74
Coarse and fine diffraction gratings

Apparatus

1 lamp, 12 V, 36 W†	item 73
or 1 compact light source†	21
1 lampholder on base	74
1 transformer or L.T. variable voltage supply	27(59)
32 (or 16) coarse diffraction gratings‡	191/1
32 (or 16) fine diffraction gratings‡	191/2
1 retort stand, boss, and clamp	503–506
32 (or 16) red colour filters	192/1
32 (or 16) green colour filters	192/2

† Mount the light source high up at one end of the laboratory. The bright compact light source will make the many spectra

Procedure

Pupils are asked to look through the coarse grating at a small bright light source far away. They are asked why the fringes do not look so close together as in the experiment with two slits. They are also asked to place first a red and then a green filter between the grating and their eyes. 'What does that tell you about those colours of light'? Then they try the fine grating.

A diffraction grating spectrum projected

Give a demonstration of spectra made by white light passing through a coarse grating and formed on a screen, then show a fine grating. This demonstration should come either just before or immediately after the class experiment. Preceding the class experiment, it makes an easier introduction for some pupils, because the spectra ('wide fringes') appear on a distant screen, as Young's fringes did. On the other hand, if this demonstration *follows* the class experiment we avoid spoiling the novelty and surprise of pupils' first look through their own grating. So we advise the class experiment first.

Demonstration 75
Spectra projected on a screen

Apparatus

1 coarse diffraction grating	item 191/1
1 fine diffraction grating	191/2
1 compact light source	21
1 L.T. variable voltage supply	59
1 cylindrical lens (+7 D)	94H
or 1 spherical lens (+7 D)	[112]
2 lens holders	124/1
1 red filter	192/1
1 green filter	192/2
2 retort stands and bosses	503–505
long white screen or wall	

easy to see, but they will not be tall. The line filament lamp will make each spectrum a taller band, so it may be better preparation for the exhibit of spectra where it will certainly be needed.

‡ The coarse grating should have 500 to 1000 lines per cm. The fine grating should have about 3000 lines per cm. Neither grating should be blazed.

Procedure

Arrange the compact light source at one end of the darkened lab and place a lens in front of it to form an image of the filament on a white screen or wall at the other end. The screen should be long, three metres or more, made from a roll of lining paper.

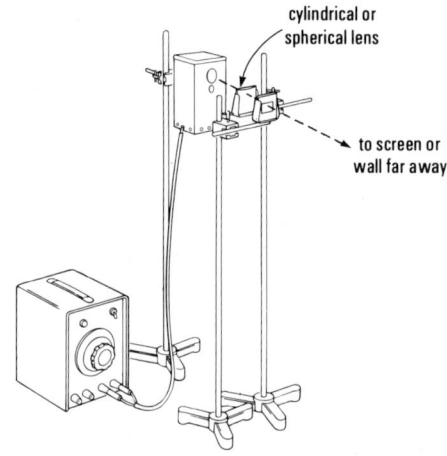

cylindrical or spherical lens

to screen or wall far away

Place the coarse diffraction grating just beyond the lens. Pupils look at the pattern and compare it with what they saw for Young's fringes.

Hold a green filter in the beam of light, then switch quickly to a red one.

Ask what pupils would expect to see if the grating could be changed to one with its furrows (slits) three times closer together. (This is worth a little time for thinking and guessing.) *Then* change to the fine grating.

Note Since the gratings only spread the spectra in one direction – horizontally, if the rulings are vertical – it may be better to use a *cylindrical* lens (+ 7 D) with its axis parallel to the rulings. This will give taller spectra although an ordinary lens (+ 7 D) makes them look brighter. The choice is a matter of taste.

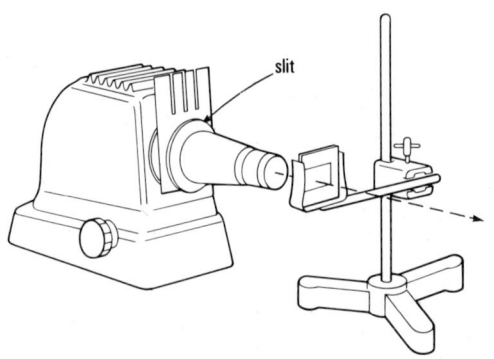

slit

Alternative arrangement with a projector. A slide projector can be used instead of the compact light source. Insert in the slide holder a card with a narrow slit cut in it. Focus the projector to form an image of the slit on the remote screen. Place the grating in front of the projector. In some cases this is easier to set up. The first method is more open for pupils to see what is happening, so it is preferable.

Demonstration 76
Ripple tank with a 'grating' of slits

Apparatus

1 ripple tank	item 90
1 lamp	57
1 vibrating bar	90L
1 transformer	27
2 dry cells for motor	52B
1 rheostat (10–15 Ω)	541/1
2 large barriers	90D
6 small barriers	90E
32 hand stroboscopes	105/1

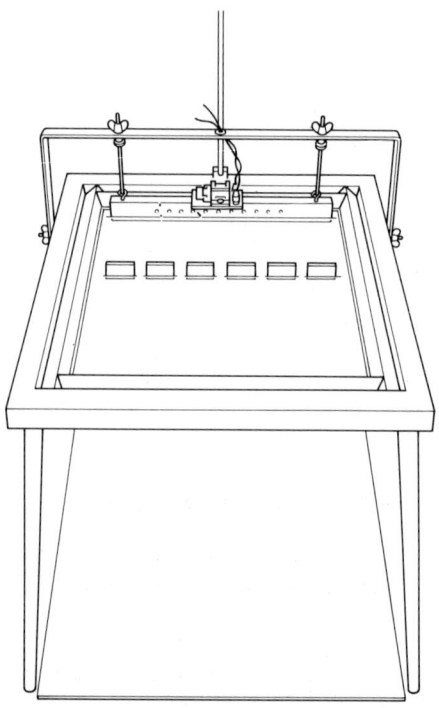

Procedure

a. Set up the ripple tank with water about 1 cm deep. Run the vibrating bar to make straight ripples. Place two long barriers 2 to 3 cm apart and about 5 cm from the vibrator.

Start the motor at a low frequency (about 4 rev./second). Pupils watch the wave passing through the gap. Increase the speed of the motor. Avoid high speeds which cause unwanted vibrations. Keep the slit width constant, at least for this first experiment.

b. Replace the long barriers by a line of short barriers with gaps of 2 or 3 cm between.

Start the motor at a low speed (4 rev./second). Pupils watch, with and without a stroboscope.

Then increase the motor speed. Pupils watch the changes in pattern.

Where a motor stroboscope is available, it may be used with advantage.

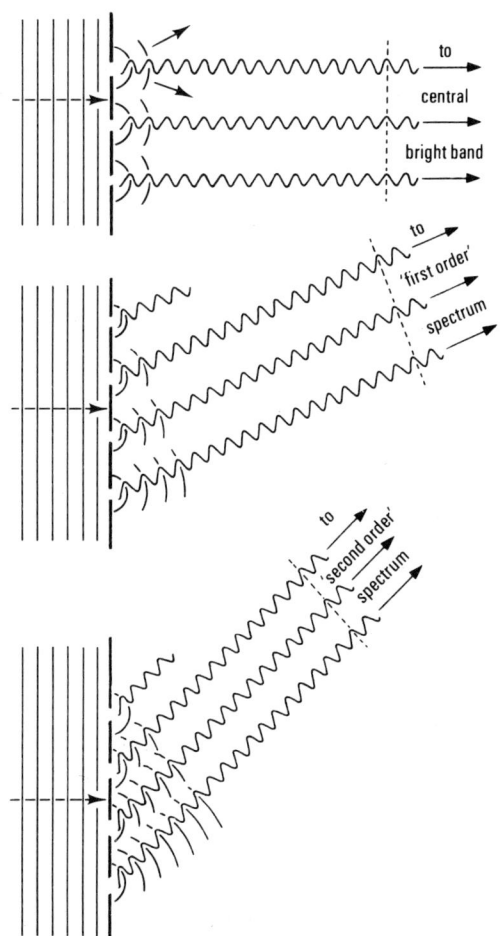

Film 76X
Ripple tank showing grating diffraction

As a possible alternative, show a film restricted to this effect.★

EXPLANATION OF THE GRATING'S ACTION

Now or earlier make sure that pupils understand how the rulings send out wavelets which contribute to first, second, . . . order spectra by arriving in phase (see diagram, above right).

From now on we should change from saying 'slits' to saying *'rulings'*. And we should call d the *grating space*.

ESTIMATE OF WAVELENGTH

This should be a class experiment. It means much more to a pupil to conduct his own measurement of

this tiny, almost inaccessible distance from crest to crest of a light wave than to see someone else do it and make a quick calculation. This leads only too easily to a formula-dominated examination question.

We hope that some pupils will be able to take a piece of grating home★ and use it as a sequel to their home experiment with Young's fringes.

★ In Year 3 we urged teachers not to show ripple tank films because our class experiments with ripples are *not* intended to provide factual knowledge in completely correct form. They

are intended to give pupils an opportunity to work at physics on their own, for the experience of experimenting.

Some pupils will emerge from ripple-tank work with definite rules of wave behaviour; others will emerge with only a general memory of having done their own experiments. And with those experiments such a range of yields will not matter. But, then to show films 'to put things right' would threaten to upset a pupil's sense of pride in depending on what he has done on his own.

Now in Year 5, a film of refraction and one of diffraction of ripples by a grating could give welcome help.

★ See the footnote on the J. Willmer Trust Fund on p. 115.

Class Experiment 77
Measuring the wavelength of light

Apparatus

1 12-V, 36-W lamp	item 73
1 lampholder (S.B.C.) on base	74
1 L.T. variable voltage supply	59
1 green filter	192/2
16 fine diffraction gratings	191/2
32 metre rules	501

Procedure

Pupils work in pairs.

Set up the 12-V 36-W line-filament lamp at one end of the laboratory, high up, so that pupils can see clearly. For a large class, set up a lamp at each end of the room, so that half the class can work facing one way, with half facing the other way, each pair as far as possible from their lamp. Place a green filter in front of the lamp. If necessary increase the applied voltage to 14 or 15 V so that there is enough light.

Pupils follow these instructions.

★　　★　　★　　★　　★

Hold a metre rule straight out in front of you towards the lamp, with the near end of the rule at your face. Hold the diffraction grating against the near end of the metre rule and look at the lamp through it.

Ask your partner to place another metre rule, at 90° to your metre rule at its far end (see the sketch). He should hold a pencil vertically above *his* metre rule and move it along until you see it in the green region of your bright spectrum. Record the distance, *x*, along your partner's ruler from the pencil to the far end of *your* ruler.

When you have made your observation, record it and change places with your partner so that he can take his turn.

Divide your measurement *x* by the length of your ruler, 100 cm. This gives you tan *A* where *A* is the angle between the line of direct white light and the light to the green in the spectrum marked by the pencil.

From tan *A* find the angle *A* from tables, and thence find sin *A* from tables.

If you are not used to trig tables, make a scale drawing instead on a large sheet of paper. Draw lines to represent the two rulers, 100 cm and *x* in cm long (to scale). Complete the triangle and measure the long sloping side, *z*. Then calculate sin *A*, which you will need for your wavelength calculation, as follows:

$$\sin A = \frac{x}{\text{slanting side } z}$$

Use the formula $d \sin A =$ wavelength to calculate the wavelength of green light. You will need

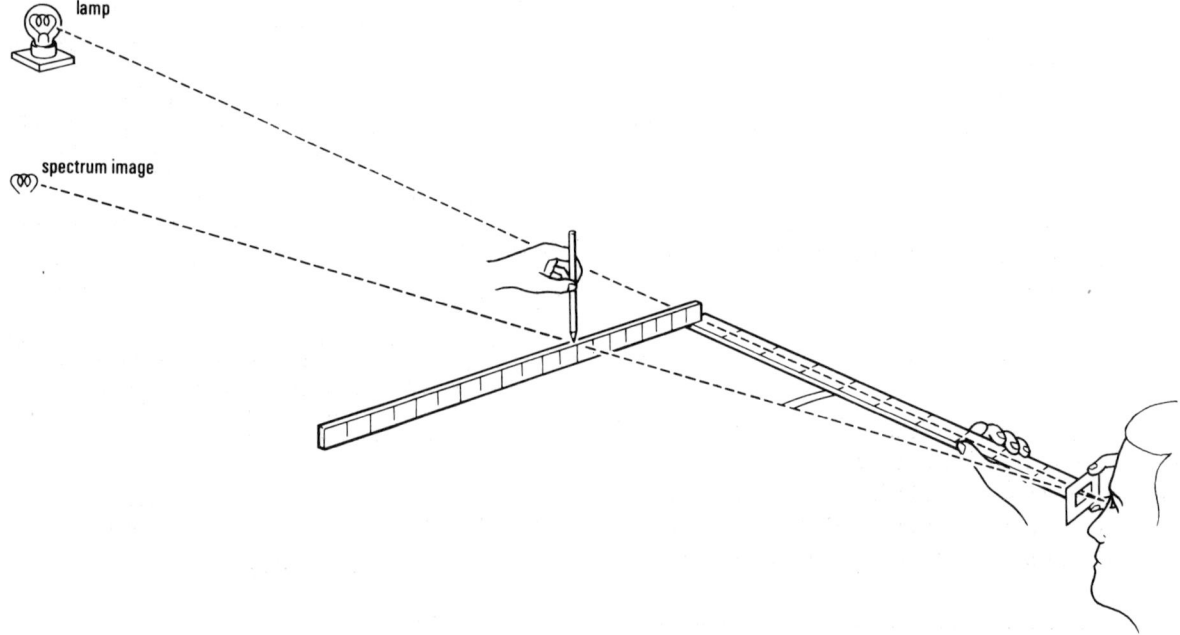

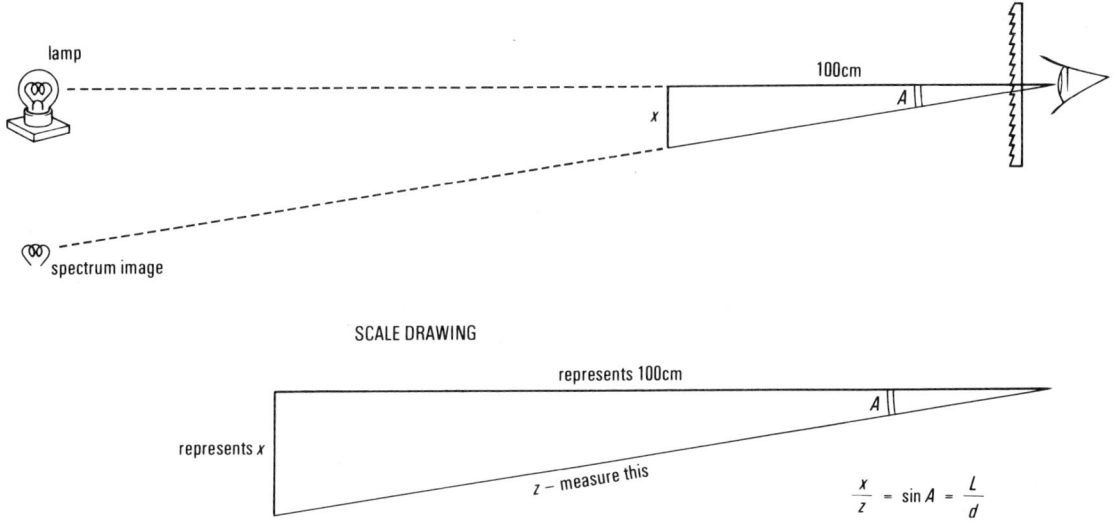

lamp

100cm

A

x

spectrum image

SCALE DRAWING

represents 100cm

A

represents x

z – measure this

$$\frac{x}{z} = \sin A = \frac{L}{d}$$

the value of d, the grating space from one ruling to the next. There are usually 3000 rulings to the cm. If so, the grating space, from ruling to ruling, is $1/3000$ cm or $1/300000$ m. Assume this, unless you are given a different value, and use it to estimate the wavelength.

★   ★   ★   ★   ★

If we simply announced the value of d, it would be almost as bad as announcing the wavelength – it would make the experiment seem obedient rather than searching. So, when we supply the value of d, we must explain where it came from and make it clear that a mechanical counting during manufacture can supply it.

If suitable microscopes are available, pupils should use them to look at their piece of grating and at the graduations on a finely divided ruler. (The $\frac{1}{2}$-mm scales provided for the oil-film experiment in Year 1 might do well.) Although pupils may not be able to measure the grating space, they will certainly see that a direct measurement *could* be made.

If this estimate of the wavelength of light will be a burden of strange geometry and unsure measurements, it would be better to omit it. If it will give a sense of delight and insight, it could be one of the most powerful experiments of the Year.

Class Experiment 78
Spectra

Apparatus

32 (or 16) fine diffraction gratings	item 191/2
1 lamp (12 V 36 W)	73
1 lampholder (S.B.C.) on base	74
1 L.T. variable voltage supply	59
32 (or 16) red filters	192/1
32 (or 16) green filters	192/2
1 neon spectrum tube†	193/1
1 hydrogen spectrum tube‡	193/2
2 holders for spectrum tubes	194
1 E.H.T. supply	14
1 Bunsen burner, boss, and stand, for sodium flame§	503–505, 508

† The spectrum tubes will require a holder and a voltage supply. Some manufacturers supply a special holder with built in voltage supply. Alternatively, a simple holder can be used with the E.H.T. power supply to give the necessary voltage. Include a resistor in series: about 1.5 MΩ (at least 2 W rating.) A p.d. of $2\frac{1}{2}$ to $3\frac{1}{2}$ kV is needed to trigger the hydrogen tube and the discharge current is about $\frac{1}{2}$ mA.

‡ A tube with hydrogen at a density to show the *atomic* spectrum is apt to develop the molecular spectrum in the course of use.

§ A sodium flame can be made by coating a piece of iron wire with sodium chloride or sodium bicarbonate; heating it in a clear flame is sufficiently bright for this experiment. A sodium pencil is a good alternative to the coated wire. A 'special' sodium lamp is not necessary.

Procedure

Pupils follow these instructions.

★   ★   ★   ★   ★

125

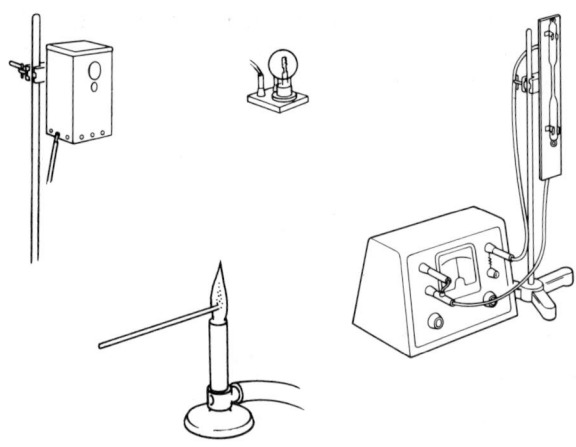

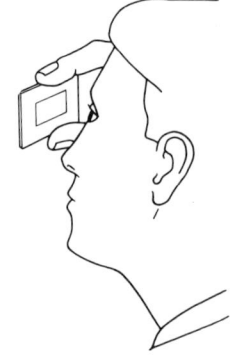

(*i*) Look at the white-hot filament of a lamp with the fine grating held close to your eye. The grating has about 3000 slits across every cm of width. You see a central white line where waves of all colours go straight through the grating. Out to each side, you see a wide bright band, which corresponds to the first bright fringe out from the centre of Young's fringes (one wavelength path difference). Since the light is white, each bright fringe is spread into a wide spectrum of colours.

Looking further out to each side, you may see a still wider, but fainter, spectrum, which corresponds to the next bright fringe out from the centre (two wavelengths path difference).

(*ii*) Look through your grating at the neon source. It is a tube containing neon atoms which are being bombarded by electrons, etc., driven by a high voltage. The red light comes from neon atoms as they recover from that excitation.

Using your grating as a spectroscope – an instrument to spread coloured light out into a spectrum – you can see that the neon atoms do not all give out red light alone. They are not behaving like a musical instrument playing a pure musical note. They are playing like a huge well-organized orchestra.

(*iii*) Now look at the hydrogen tube. There, hydrogen atoms are being bombarded. As they recover, they give out only a few definite colours. They behave like a much simpler orchestra: so much simpler that the hydrogen spectrum was one of the first to have its vibrations decoded and so yield very important information about atomic structure.

(*iv*) Look through your grating at a yellow sodium flame, a clear flame with salt added. Do you agree that it gives out pure yellow light, not the usual mixture of red and green which we accept for 'common' yellow in colour mixing?

★ ★ ★ ★ ★

ELECTROMAGNETIC SPECTRUM

Pupils' Text gives notes, and shows a long chart of frequencies.

SPECTRA – OPTIONAL DEMONSTRATIONS

Spectra are beautiful things to see. The full spectrum of white light shining on a screen in a large demonstration is a surprise and a delight to pupils, even though they have seen a dilute version often enough in a rainbow, and probably a smaller version with their ray streaks. Teachers may wish to show the spectrum produced by a prism.

Demonstration 78X
Spectrum (*OPTIONAL***)**

Apparatus

1 high dispersion prism†	item 69
1 convex lens	113/3
1 lens holder	124/1
1 compact light source	21
1 L.T. variable voltage supply	59
2 retort stands and bosses	503–505

† If possible, with *aperture* 30 mm or more.

Procedure

Set up the source at one end of the lab. (If a 12-V 36-W lamp is used instead of the compact light source, overrun it at 14–16 V.)

Arrange the lens to form an image of the source on a distant screen. Then place the prism just beyond the lens, and move the screen to catch the spectrum.

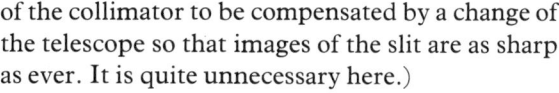

of the collimator to be compensated by a change of the telescope so that images of the slit are as sharp as ever. It is quite unnecessary here.)

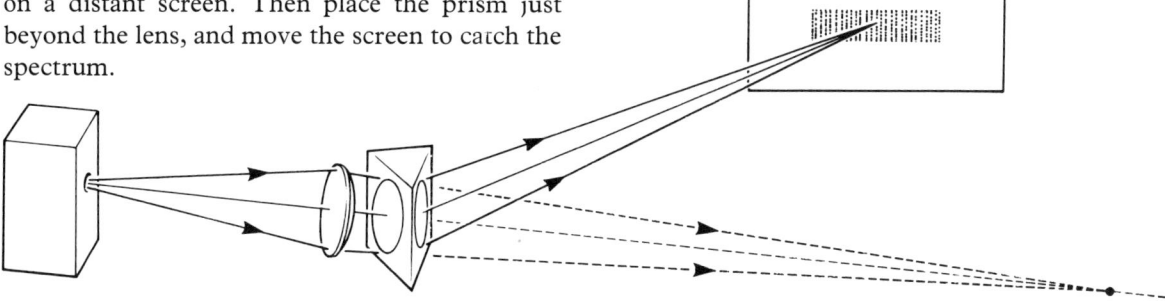

Turn the prism to make a wide spectrum. Also turn the screen to receive the spectrum obliquely, if that suits the arrangement of the room. See the note in the box for commentary on the prism.

Point out the advantage of such a spectrum: good for observing, though more difficult (or, rather, indirect) for measuring.

Note The filament of the compact light source is short as well as narrow, so the spectrum will look more like a *line* of colours than a tall band.* For a tall band, use a lamp with a longer vertical filament.

Note for spectrum demonstrations

A long spectrum is much more impressive than a short one – in fact, it seems to make a different impression (like the difference between a coloured dress and a coloured ribbon).

A bright, long spectrum can be made by meeting a combination of five conditions:

(*i*) Use a high-dispersion (flint glass) prism – well worth the cost.

(*ii*) Turn the prism to give a much longer spectrum than at a minimum deviation. (Minimum deviation is only a counsel of perfection in precise spectroscopy, where it enables slight mis-focusing

(*iii*) Turn the screen to receive the spectrum at a slant, thus spreading it.

(*iv*) Use a large 'aperture'. For the lens alone this would mean a large *f*-number, but the prism makes the practical limitation. We need a squat prism, no taller than the height of the incident beam of light (more than that is pure waste) and wide enough to make use of the full width of the incident beam. Since the prism receives the incident beam obliquely, the width of its face should be much greater than the width of the beam. For example, a high-dispersion prism that can take in a beam 32 mm wide (at minimum deviation) *must have a face more than 50 mm wide*. A sensible specification of such a prism could be '*aperture* 32 mm' – in contrast with a prism of '*face width* 32 mm', which would miss 35 per cent of the light!

(*v*) Then, since all the light that enters the prism is distributed over the area of the spectrum, a strong source is needed. We recommend no slit but a vertical line filament. The compact light source is the best of our suggested equipment, though its spectrum is not tall. A 12-V line filament lamp run at higher voltage is good. A tungsten-halogen lamp with a longer filament, bought for this purpose and set in a special housing would be best of all.

OTHER SPECTRA

Line spectra are interesting things to see, but they are not as interesting to pupils as to us, and they certainly do not present themselves to pupils as keys to atomic structure.

Pupils' Text 5 includes a comparative set of

---

* This can be avoided by inserting a *cylindrical* lens. Teachers may find it interesting to experiment in modifying a spectrum with cylindrical lenses from the Year 3 ray-streaks apparatus; but in general this adds complications and takes time, so we do not recommend it as a standard arrangement.

photographs of line spectra. The sensitivity of the film used in making these photographs dropped rapidly in the near infra-red. This accounts for the cut-off at the red end of the spectrum of the tungsten filament lamp, and for the partial suppression of the Hα line.

An absorption spectrum, such as that given by a green filter held in a beam of white light, helps pupils to understand the physics of colour mixing.

We do not in general advise using focused spectroscopes with pupils at this stage. But the trouble of arranging a spectroscope for pupils to see the Sun's spectrum with fine resolution is well worth while.

The 'single line' spectrum from a salted flame is a surprise: instead of a broad band of yellow, pupils see a very narrow band of pure yellow. With a fast group, the *absorption* spectrum of sodium vapour then provides an interpretation of the Sun's absorption lines.

Notes on experimental procedures follow.

Demonstration 78Y
Absorption lines in the Sun's spectrum
(*OPTIONAL*)

Apparatus

1 spectrometer
1 prism (*preferably high dispersion*)   item 69
1 positive lens (+2·5D)   113/3
1 lens holder   124/1
plane mirror

Procedure

Set up a prism spectrometer in the usual way to give a good spectrum of white light.

With the slit very narrow, direct sunlight into it with a plane mirror. Use a lens to converge the light to a focus, a few cm in front of the slit.

Arrange this experiment in a partially darkened room so that the spectrum will be sufficiently bright. Details must depend on the layout of the lab, and are likely to require careful planning.

Make sure that the sunlight continues to fall on the slit and travel through to the prism.

If the dark lines do not appear, *make the slit narrower*.

Pupils view the spectrum one at a time, through the spectrometer's telescope. Do not tell them to look for absorption lines, but simply say, 'Look at the spectrum. Is it brighter than usual at the blue end?' Then when pupils see the thin black lines it is time for some explanation.

That will practically necessitate a demonstration to show the way in which such absorption lines can be produced. Though this may seem an advanced extension of our work that hardly fits here, we include it because of its great importance in astrophysics. Such an absorption spectrum tells us the composition of the outer layers of incandescent stars, like the Sun. Slight shifts in the position of the lines probably tell us the speed with which stars are approaching or receding.

Demonstration 78Z
Absorption spectrum of sodium
(*OPTIONAL*)

Apparatus

1 lamp (12 V 24 W)	item	72
1 lampholder (S.B.C.) on base		74
1 L.T. variable voltage supply		59
2 positive lenses (+7D)		112
2 lens holders		124/1
2 retort stands, bosses, and clamps		503–506
1 spectrometer		
1 prism (preferably high dispersion)		
Bunsen burner with salt for sodium flame		

Preparation

It is essential to provide an intense sodium flame. An iron wire or a ceramic rod dipped in concentrated brine and held in a Bunsen flame will provide this.

Focus the spectrometer in the usual way. Keep the slit narrow.

Procedure

Direct white light from the line filament lamp to the slit. Arrange the two positive lenses to make

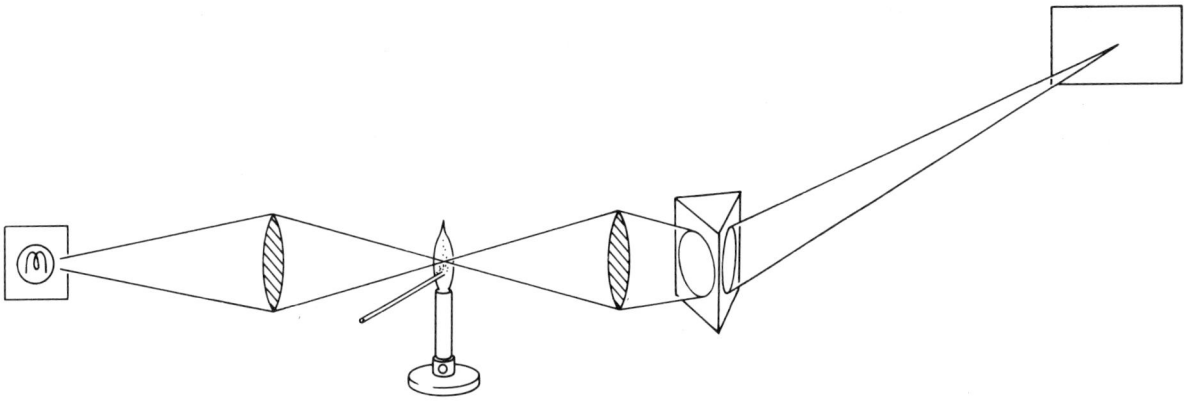

sure that *all* the white light entering the collimator has passed through the flame on the way to it.

One lens forms a real image of the filament *in the flame*; the other lens forms a real image of that first real image *on the slit* of the spectroscope. It is easiest to make each of the distances filament to lens, lens to image, image to lens, and lens to slit, twice the focal length of the lens concerned.

Place the intense sodium flame mid-way between the two lenses, where the first lens produces a sharp image. This makes all the white light pass through the flame in the region where it is rich in sodium.

Adjust the voltage applied to the lamp, as well as the slit width, to get the best conditions for seeing the dark lines.

Pupils may also see the picture change to the yellow emission spectrum of sodium, if the white light is dimmed.

CHAPTER 10

Radioactivity: Experimental study, nuclear atom neutrons, reactors

This should be an experimental study using a combination of class experiments and demonstrations. We suggest that the emphasis should be on the effects of radiations in making ions, on exponential decay, and on nuclear changes.

Since radioactivity is a topic of considerable interest teachers may feel tempted to spend a long time on the section. But since we hope there will be enough time to proceed to the photoelectric effect, theories of light, and matter waves to carry forward our building of atomic models, we suggest that the teaching of radioactivity should not be prolonged. Pupils who are specially interested will find that there are good books. For example, an excellent account of the early history is given in *The restless atom* by A. Romer (Heinemann Science Study series).

ATOM MODELS SO FAR

In describing solids, liquids, and gases, and in developing kinetic theory, we treated molecules and atoms as round knobs with no internal structure. We described them as exerting attractions on each other at short ranges of approach and – necessarily – repulsions at very short range. We assumed that those forces are 'the same on the way in as on the way out' when one atom or molecule approaches another; so that collisions and other interactions are elastic.

Then in Year 4, pupils met inelastic effects, when electrons are torn off atoms by violent electric fields, or knocked off atoms in violent collisions. The idea of making ions in gases as well as in solutions was important to the understanding of Millikan's experiment.

We illustrated that idea by showing currents being driven through air by an electric field when a candle flame was placed in it, when a lighted match provided ions, and when a radioactive source was held nearby.

These currents are too small to demonstrate with a microammeter, so we use an electroscope and show the leaf rising as charge is driven across to it, or falling as charge leaks away without any driving battery. Such ions can only be produced at the expense of some supply of energy to drag electrons off uncharged atoms. We did not say that then, but we should say it now.

In Year 4 we showed some properties of electron streams released by a hot filament in a vacuum. Now in Year 5 pupils have measured e/m for the particles in such streams. And, comparing the result with the value for hydrogen ions in solution, they should picture electrons as very small chips off atoms.

Thus, by now, pupils should have a picture of atoms as containing electrons (some or all fairly easily detached) and some positively charged material, probably holding most of the mass of the atom, the whole being held together in an unknown way.

IONS IN AIR CARRY CURRENT

Remind pupils of this tentative atom model – electrons embedded in positively charged material. Ask what ways they know of removing electrons to make ions. Perhaps show again:

a. A candle flame, projected by shadowing, in a strong electric field between two vertical plates.

b. A high voltage arranged to drive ions across an air gap to an electroscope. And, with the high voltage removed, ions carrying charge away from the charged electroscope. Provide ions for this by a small flame. A lighted match does well.

This is *not* the time to show ions carrying current with more complicated apparatus, such as a d.c. amplifier. At this stage, simplicity is best.

Demonstration 79
Ions in a candle flame

Apparatus

1 pair of metal plates with insulating handles	item 65
1 E.H.T. power supply	14
1 compact light source	21
1 candle	
white screen or wall	

Procedure

Set up the plates in vertical planes, parallel to each other, 5 to 8 cm apart, with their handles in bosses on stands.

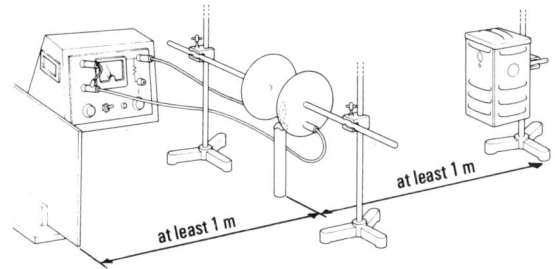

Place the lighted candle so that its flame is a little below the plates.

Place the light source a metre or more away so that a shadow of the plates and the flame falls on the distant screen.

When a high voltage is applied between the plates, the shadow of the flame and its gases above it divides into two parts, one towards the positive plate, the other towards the negative plate.

Demonstration 80
Ions carry charges to or from an electroscope

Apparatus

1 pair of metal plates with insulating handles	item 65
1 E.H.T. power supply	14
1 electroscope	51A
1 hook for electroscope	51J
2 retort stands and bosses	503–505

Procedure

Support the plates parallel to each other about 5 cm apart.

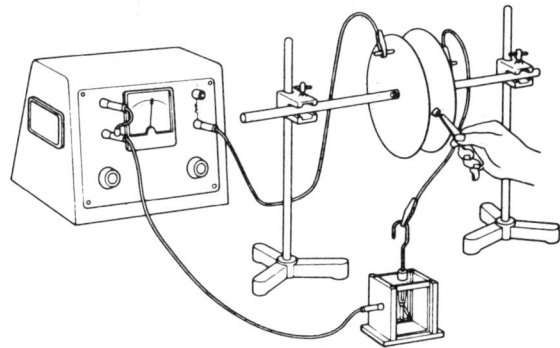

Connect one plate to the positive terminal of the E.H.T. supply through the safety resistor, the other plate to the leaf of the electroscope through the hook. Connect the case of the electroscope to the negative terminal of the E.H.T. supply, which is itself earthed.

a. Apply about 2 kV to the plates. The electroscope leaf will show induced charges, and it should be earthed momentarily to remove them.

Hold a match flame just below the plates. Pupils watch the leaf.

b. Show ions discharging the electroscope. The power supply is no longer needed: the electric field between the electroscope's leaf and case suffices.

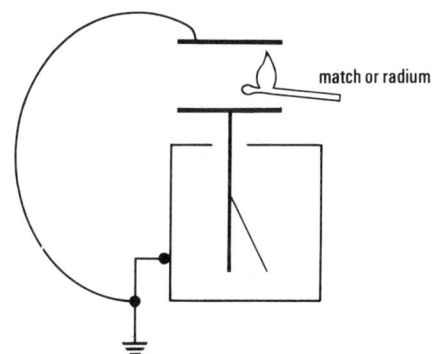

match or radium

Remove the high-voltage connections, and show one of the plates connected to the electroscope leaf and the other connected directly to the earthed case. Charge the electroscope using a flying lead from the power supply, or using a charged rod. Hold the match flame under the space between the plates.

Note Pupils will see the leaf more easily if the front of the electroscope is covered with translucent paper and a shadow of the leaf is cast on it by a 12-V, 24-W lamp behind the electroscope. With a large class it may be better still to cast a shadow of the leaf on a distant wall.

Radioactivity Now pupils are ready to meet radioactivity with more understanding. Radioactivity was discovered by the ionizing effect of the radiations from certain substances, and at first that was its chief property. Nowadays we are even more interested in the nuclear changes that release the particles which produce that ionization; but pupils should start by seeing the original property.

Repeat the last experiment with a small radium source instead of the match flame.

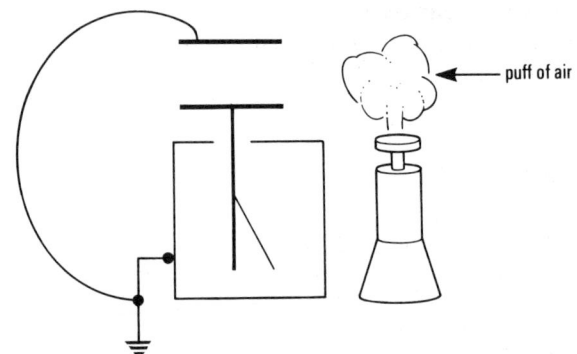

Demonstration 81a
Radioactive material makes ions in air

Apparatus

1 electroscope and plate	item 51A, B
(see Note for **Demonstration 80**)	
1 plate with insulating handle	65
1 5 µCi radium source	16
1 source holder	196
1 retort stand and boss	503–505
1 E.H.T. power supply	14

Procedure

Clamp the plate by its insulating handle so that it is held horizontally 2–3 cm above the cap of the electroscope and connect it to the case of the electroscope. Use a flying lead from the power supply to charge the electroscope.

Then bring the radium source in its holder near to the cap. The initial fall of the leaf is due to the change in capacitance of the system because the hand is an earthed conductor. But subsequent falls with the source held quite still are due to ionization. The charged electroscope's own field drives ions, which thus carry charge across.

Demonstration 81b
A current of air can carry ions

Repeat the last experiment with the charged electroscope but this time hold the source of ions out at one side, pointing upwards as in the sketch.

Charge the electroscope. Pupils watch the leaf: it will hardly move, because ions are not getting into the space between the plates.

Then puff some air across, above the source and into the space between the plates. Pupils watch the leaf as this is done. Try both the match flame and the radium source for this.

PREPARATION FOR UNDERSTANDING GEIGER COUNTERS

Pupils will understand our demonstrations of radioactivity much better if the Geiger counter is not presented suddenly as a 'black box'. They may have seen it used without much explanation in Year 1; and they have seen the scaler used as a clock in Year 4; but we should now show some experiments that will help pupils to consider a GM tube a sensible device. If the experiments described below seem to teachers rather childish, we hope they will try the treatment once, because we have found it helpful in introducing the counting of ionizing particles.

Introduction to Geiger counters The next two demonstrations are intended to lead up to an explanation of Geiger counters, in a light-hearted way. To get the best from them, introduce each as a mystery (as in *Pupils' Text*) without any reference to Geiger counters; show the experiment: and only then give the joking name. After both are done, offer 'one more mystery', with the help of a radioactive source. Show the spark counter and give that its proper name. *Pupils' Text* describes the mechanism of each of the three, but omits the joking name.

Demonstration 82a
'Mystery experiment I' (later to be called the 'salt counter')

Apparatus

1 12-V 36-W lamp	item 73
1 S.B.C. lampholder on base	74
1 L.T. variable voltage supply	59
2 retort stands, bosses, and clamps	503–506
1 beaker, 1000 cm^3	
2 lengths stout copper wire	
Supply of table salt	
Distilled or de-ionized water†	

† Most tap water already conducts too well.

Procedure

Set up a series circuit of the lamp, the power supply, and the two copper wires. Let the bare ends of the wires touch. Switch on the 12-V supply. The lamp lights.

Switch off; hang the wires in the beaker well apart. Switch on again. No effect visible – 'ordinary air does not conduct'. Switch off.

Pour distilled water into the beaker to cover the bare ends of the wires. Switch on. No effect visible.

Tell pupils that electrically charged carriers are needed – electrons in wires or ions in gases or liquids. Salt is made up of ions Na^+ and Cl^- (already separated, even in the solid crystal).

Throw a handful of salt into the water. The lamp lights.

Remind pupils that the p.d. between the wires makes an electric field across the beaker, through the water. If there are charged particles there the field will tug positive charges one way, negative charges the opposite way; the charges will move and that will be a current. When the salt went in it must have provided some charged particles to act as carriers – ions.

Say, as a joke – unexplained for the moment –

that this is not just another demonstration of electrolysis. This is a 'salt counter' for counting handfuls of salt. ('I throw in the handfuls, you count the times the lamp lights. Of course the beaker would have to be washed out and refilled after each count.')

Demonstration 82b
'Mystery experiment II' (later to be called the 'match counter')

Apparatus

1 E.H.T. power supply	item 14
2 insulated metal balls	188
2 retort stands, bosses, and clamps	503–506
1 capacitor (0·001 µF, 20 kV)	132E

Procedure

Set up the two balls a centimetre or so apart, holding their insulated handles with bosses on retort stands.

Connect the large capacitor, well insulated from earth, across the gap. It will make the spark fat and noisy.

In a rehearsal beforehand, adjust the separation between the balls; find the optimum distance, when the arrangement just fails to spark spontaneously.

Apply about 5000 volts as follows: connect one ball to the earthed negative terminal of the E.H.T. power supply. Use a well insulated flying lead to make momentary connection between the other ball and the positive terminal of the power supply.

Explain that the supply will drive + and – charges to the balls and thus make a very strong

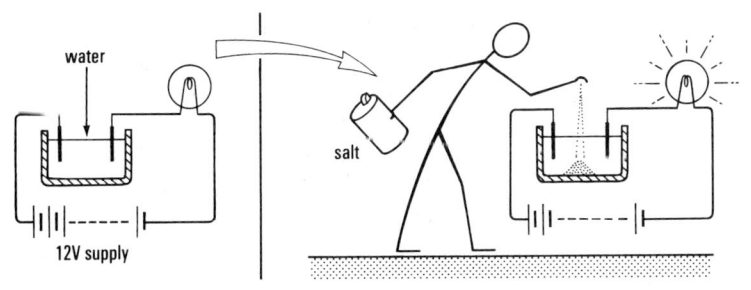

electric field between them. Positive and negative charges are stored up not only on the balls but also on the plates of the big capacitor, so that there are big charges waiting to be driven across the gap. However the electric field between the balls is not quite big enough to start a spark. 'But the camel's back is loaded very strongly. It will only take one or two straws to break the camel's back; it will only take a few ions in the air in the gap to start a spark. Now watch.'

When the balls and capacitor are charged, hold a lighted match just under the gap. The ions in the flame will trigger off the spark.

In the spark, ions driven by the electric field make more ions in collisions – there is a chain reaction, an avalanche of electrons and ions.

Say, as a joke – unexplained for the moment – that this is not just a demonstration of sparks, but a 'match counter' ('*I* light the matches, *you* count the sparks').

A spark is due to an avalanche of electrons, a chain reaction of ionization by collisions. Pupils already know that a flame provides enough ions to start the avalanche if the electric field is strong enough. Tell them: 'In that spark, we were not just driving a few ions across put there by the flame. They were only the pacers for the full team running the race.'

Demonstration 82c
'Mystery experiment III' (soon to be called the 'spark counter')

Apparatus

1 spark counter†	item 17
1 E.H.T. power supply	14
1 radium source	16
1 source holder	196

† The new form, now available, has many wires, with no obstructive screen. It is safe and is better for teaching. (See the description below.)

Procedure

Connect the positive terminal of the E.H.T. power supply to earth (without the 50-MΩ safety

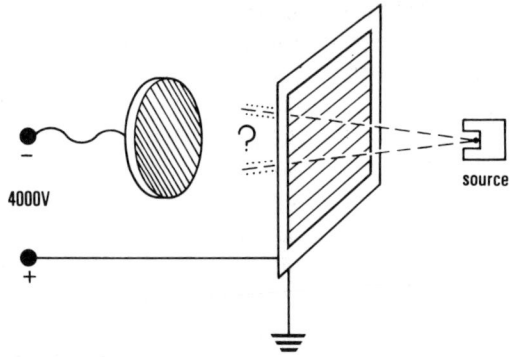

resistor), and to the wire grid of the spark counter. Connect the back plate of the counter to the negative terminal of the supply.

Turn up the voltage *slowly* until it is just below the point of starting a spark. Usually a p.d. of about 4500 volts is necessary.

Show the grid of wires and the plate behind. Explain that once again the voltage between plate and wires is almost great enough to start a spark. This time, instead of a match flame, a small sample of radioactive material will be brought near. The nuclei of the radioactive atoms emit missiles which make many ions in the air in their short path.

Hold the source just in front of the grid. Sparks will be seen and heard, obviously occurring at random.

Move the source slowly away to show the short range of the radiation. It is a stream of alpha particles of range a few centimetres.

We might say 'Now we have an alpha particle counter. The radioactive material emits the alpha particle, you count the sparks.'

Sparks and ionization by collision Pupils need not find electric sparks mysterious – though if any of us knows the full complexity of a spark he might well feel puzzled. Rather than put a long story in *Pupils' Text* we can refer them to the account in Year 4. The *Teachers' Guide* for that year provides notes of the things one might say to an enquirer.

Description of equipment for spark counter

In the first edition, we suggested a device consisting of a single taut wire stretched above a metal plate which served as the other electrode.

134

The wire was protected by a screen of metal gauze in front of it; but this did not make it easy for pupils to see the sparks. Also, the radioactive source was usually carried along by hand over the wire, so the series of sparks seemed more like a record of the motion of the source than an exhibit of random events in time. An open form can now be obtained. It is so much better for teaching that we hope schools will buy or make one and substitute it for the older form.

GRID The new device consists of a grid of very fine wires that are taut and parallel, a few millimetres apart. The grid is held about two millimetres in front of a metal plate which is the other electrode.

The wires are supported on a square frame of metal. Since they must all be kept taut it is best to use a single wire stretched to and fro between pegs on the frame. The wire must be tungsten or steel of very fine gauge so that when it is charged the electric field near it is very large, enough to make a chain reaction of ionization by collision in air at 4 kV. The wires must be *positively* charged to receive an avalanche of electrons when a spark is started.

Both electrodes, the frame of wires, and the solid disk, should be mounted on thick pieces of perspex with three spring-loaded screws to adjust the distance apart and to make sure that the plane of the grid is parallel to the plate. The use of perspex makes the whole arrangement visible to pupils – the device is simple, and it would be a greaty pity to give it an unnecessary 'black box' appearance.

When an alpha-particle source is brought near the grid, small bright sparks jump between the wires and the plate with a random distribution in time, and in a random pattern of places.

THE POWER SUPPLY must be adjustable to give a p.d. of 3000 to 5000 volts between the grid and the plate. When the device is to be used, turn up the voltage very slowly until the electric field between grid and plate reaches a value just short of that which would start sparks spontaneously. Then an alpha particle flying in between the wires of the grid will make enough ions to start a spark.

The power supply must provide *well smoothed and regulated* d.c.; the pulses of plain rectified a.c. would not be suitable.

CAPACITOR Each spark must be quenched quickly. This is ensured by a small capacitor connected between the electrodes. The capacitor is charged by the power supply but when there is a spark it discharges, momentarily, through the spark; the voltage between the electrodes drops and the spark stops. The larger the capacitance the brighter and louder is the spark, but also the longer is the dead time before the counter is ready for another spark. A capacitance of about 1000 picofarad is suitable: the sparks will be visible and audible over the room yet the dead time will be short. (Some power supplies have a capacitor inside, across their output, and that may suffice.)

Since with the E.H.T. power supplies suggested for this programme either terminal can be connected to earth, the device is quite safe in use without any special protection. The positive terminal of the E.H.T. supply should be connected to earth as well as to the grid, then the grid is at earth potential, and a hand holding a source close in front is in no danger. The negative plate behind the grid is well protected by the general perspex frame.

STOPPING Hold a sheet of paper between the source and the grid to show that it stops alpha particles from radium. A sheet of cigarette paper is just thin enough to let them through.

SCALER This 'open air' device can be connected to a scaler so that alpha particles can be counted just as beta particles are counted with a GM tube. This obviates the need for a special solid state detector and pre-amplifier for alpha particles. The connection must be made to input terminals that exclude the scaler's own high voltage supply.

The different ionizing radiations Pupils should see that an alpha particle makes a great number of ions – over 100000 pairs in its short path of a few centimetres in air. Cloud-chamber pictures emphasize that: the water drops are almost too numerous to be seen separately. Even very energetic alpha particles are stopped by a sheet of paper, but they can get through cigarette paper.

Beta particles, which are energetic electrons, hurled out at high speed, make comparable numbers of ions in a much longer path, and they take many sheets of paper or thin metal to stop them.

Gamma rays are high energy photons, very

short wavelength X-rays. They travel through matter still more easily, seldom making a collision. When a gamma-ray photon does stop it disappears completely* and an electron is ejected from the target atom.

Pupils should see demonstrations of these differences of ionization and absorption.

Demonstration 83
Radiations and counters

Apparatus

1 scaler	item 130/1
1 holder for GM tube	130/3
1 thin window GM tube	130/5
1 gamma GM tube	130/6
radioactive sources:†	
1 gamma, cobalt 60	195/1
1 beta, strontium 90	195/2
1 alpha, americium 241	195/3
1 radium	16
1 holder for radioactive sources	196
1 spark counter (new form)	17
1 E.H.T. power supply	14

† Each of the sources listed is available with 5 microcurie strength.

Closed sources are commercially available and have approval from the Department of Education and Science as far as use in schools is concerned, provided the other conditions concerning the use of radioactive materials are followed. See D.E.S. Administrative Memorandum 2/76 'The Use of Ionizing Radiations in Educational Establishments' and the accompanying 'Notes for Guidance'.

Procedure

Present each of the four sources in turn to each of the two GM tubes and to the spark counter.

Connect the tube to the scaler through the socket on the scaler which accepts GM-tube cables. Apply the proper voltage by the scaler's built-in supply.

Connect the E.H.T. to the spark counter and adjust it slowly as in **Demonstration 82c**. (The sparks can be seen and heard, but if the spark counter is also to register on the scaler it must be connected to an entry which does not include the scaler's built-in high voltage supply.)

Place absorbers between the source and the

* There is some scattering too but that is unimportant in our present look at ionization.

detecting device. Pupils will see that paper stops alpha particles, a sheet of perspex, an exercise book, or thin aluminium stops beta particles, but thick lead or very thick aluminium is needed to reduce gamma radiation appreciably.

Notes There is no need in this O-Level teaching to show the characteristic of a GM tube – how its response depends on the voltage applied. However, teachers may want to look at the characteristic of each GM tube for their own interest. To plot that, put a source at a fixed distance from the tube. Start with the voltage well below the threshold and turn it up slowly from, say, 250 volts in 25-volt stages until counts start. Measurements of the count-rate over a range of applied voltages will give a graph like the sketch.

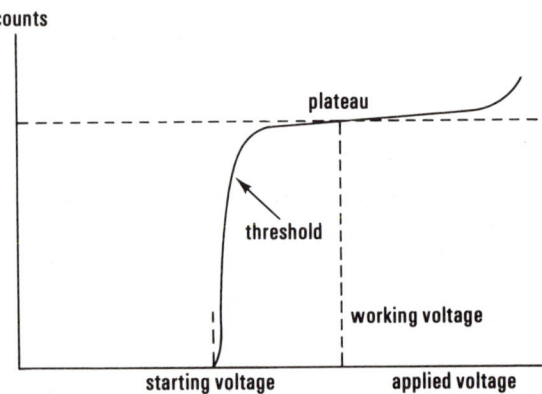

The GM tube is normally operated at a voltage on the plateau, usually 50 to 100 volts above the threshold. If the voltage is turned up even more, the count-rate will rise steeply. The tube should *not* be operated in the steep region. The test of a good tube is the length of the plateau.

The spark counter has no long plateau: its adjustment is critical.

Other forms of counter, arranged to measure a more gentle level of ionization, can measure the ionization and distinguish differences between one particle and another. For our teaching, it seems best to use tubes which just give pulses of a standard size.

2. *Scintillations* In early investigations, the tiny flashes of light made by a single alpha particle hitting a screen of zinc sulphide provided a very important tool. Counting 'scintillations' was tedious and tiring, but it provided the essential measurements of alpha-particle scattering that led to the nuclear model for atoms. Later, Geiger counters (and proportional counters) superseded that method completely for counting individual particles.

Then, decades later still, scintillation counters came back into use, with more efficient sensitive materials, and photomultiplier tubes (amplifying by millions) to take the place of human eyes.

3. *New counters* Solid-state counters are now in use, in which the impinging particle makes a similar disturbance – without a flash of light playing a part. It pushes some electrons into upper levels, leaving 'holes' that act like movable *positive* charges, in the solid semiconductor. Both the electrons and the holes are driven by a strong electric field and make a pulse of current. The pulses can be sorted by size and counted.

Demonstration 84
Experiments with alpha particles: range and stopping

Apparatus

1 alpha source	item 16 or 195/3
1 holder for radioactive sources	196
1 spark counter (new form)	17
1 E.H.T. power supply	14
1 expansion cloud chamber	18
1 H.T. power supply	15

Procedure

a. *Range* Turn up the voltage applied to the spark counter *very slowly* until sparks occur when an alpha source is held very near. Keep the voltage fixed at that value.

Present the alpha source to the spark counter at a distance of about 10 cm. Move the source slowly towards the counter until the counter responds with sparks. The positive wires of the counter are then at the end of the range of the alpha particles from the source. Repeat that approach several times to emphasize the sharply defined range.

If a spark occurs spontaneously, or if a spark started by an alpha particle is not quenched but continues, turn down the voltage immediately, to avoid damage.

b. *Stopping* Hold the source near the grid so that plenty of sparks occur. Interpose a sheet of paper between the source and grid. The sparks will stop, showing that the range of alpha particles in paper is less than the thickness of the sheet. Then try cigarette paper, which is thin enough to let some through.

c. With a small alpha source in an expansion cloud chamber the limited range of the tracks may be seen clearly.

Let pupils see, in their own experiment, the radiation from a simple radioactive source producing ionization to discharge an electroscope.

Class Experiment 85
Uranium oxide source in a gold-leaf electroscope

Apparatus

16 gold-leaf electroscopes	item 51A
16 charging hooks		51J
16 polythene strips		51G
16 rubbers		51I
16 uranium oxide sources		

Preparation

Each source should be a layer of powdered uranium oxide on a small shallow pan such as a tin lid. Paint the top of the pan with glue or waterproof cement. While the cement is still sticky powder it with uranium oxide.

Procedure

Pupils follow these instructions.

*   *   *   *   *

First charge your electroscope, as follows: attach the hook to the top of your electroscope; charge a strip of plastic (polythene) by rubbing it with a cloth. (The cloth pulls electrons off the plastic.) Scrape the charged plastic on the hook to transfer some charge to the electroscope. Then lift the hook away, using the plastic strip as an insulating handle. Without the hook your electroscope will be extra sensitive to small changes of charge on the leaf.

Watch the leaf for a minute or two. Then ask your teacher for a small tray with a thin layer of uranium oxide spread on it. The black uranium oxide powder is held on the tray with glue, to make sure it does not spill.

Lift the glass window of your electroscope and put the uranium oxide source on the floor inside it. Close the window. Charge your electroscope again and watch the leaf.

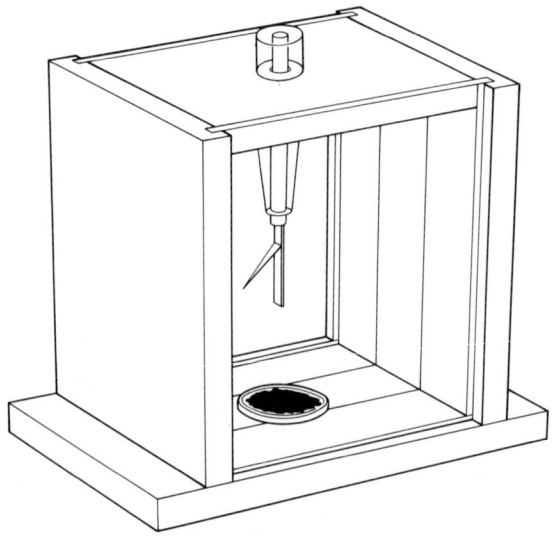

Uranium is a heavy metal. Its atoms fling out alpha particles which are not very energetic – but they make enough ions for you to see the effect.

If you like, you could compare two sources of different strengths by timing the fall of the leaf with each.

Note Although uranium oxide is not regarded as dangerous radioactive material you should obey the general rule for *all* people handling radioactive material: AFTER THE EXPERIMENT WASH YOUR HANDS BEFORE YOU EAT OR SMOKE.

⋆　　⋆　　⋆　　⋆　　⋆

Gamma rays and distance When counting gamma rays, we might show a fast group a demonstration suggesting an inverse-square law for the counting rate at various distances from a small source. Gamma rays travel, in air, on long straight paths, so we expect an inverse-square law to apply to a sheaf of gamma rays from a compact source.

Since pupils already know what an inverse-square law is like from their study of gravitation, and for illumination, they should appreciate this example. The experiment has an important moral: in safeguarding oneself from gamma radiation the best thing is to move further away. At ten times the distance in air one should feel 100 times as safe.

Understanding the cloud chamber As preparation for pupils to see cloud chambers with fuller understanding, show clouds forming when wet air expands suddenly.

Demonstration 86a
Making a cloud

Apparatus

1 large flask (or aspirator)	item 523
1 bung with glass tube and short rubber tube	
1 lamp	46/2

Procedure

Put a little water in the flask. Close the flask with the bung and tubing.

(*i*) Blow into the rubber tube to raise the pressure; pinch the tube and hold it for a short time to let the air in the flask return to room temperature. Release the pressure suddenly, preferably by pulling the bung out abruptly. A cloud will form. To make sure it is easy to see, darken the room and shine light on the flask from one side. Place a dark screen behind the flask.

Replace the bung and blow into the flask to show the cloud disappearing.

(*ii*) Make a cloud in the flask, let it settle. Repeat several times. This will remove many of the particles that start cloud droplets; then the cloud formed on expansion will be thinner. But a dose of ions and attracting particles supplied by opening the flask and throwing in a lighted match leads to a thick cloud on the next expansion.

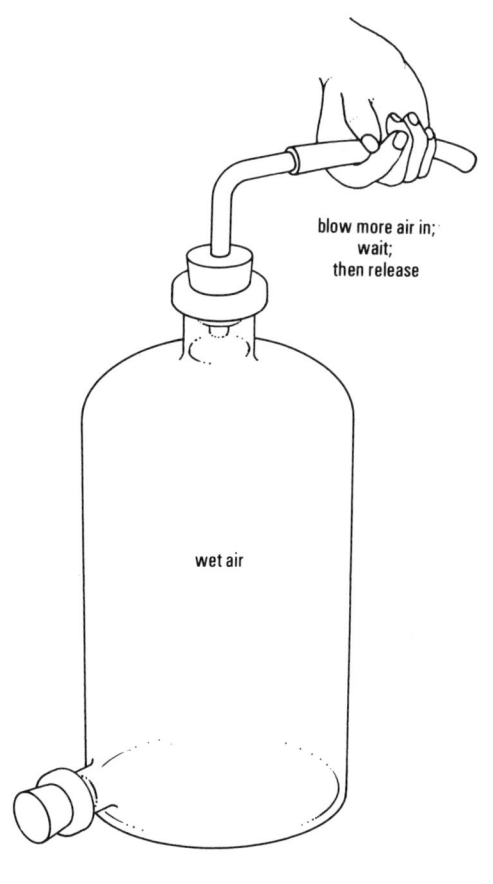

blow more air in;
wait;
then release

wet air

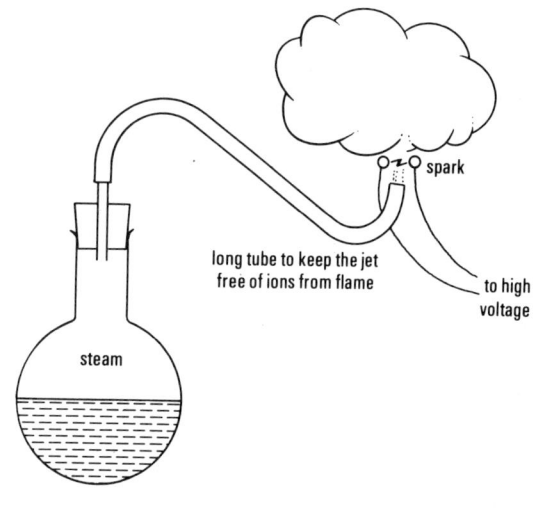

spark

long tube to keep the jet
free of ions from flame

steam

to high
voltage

1·5 metres of rubber tubing
1 Van de Graaff generator†   60/1

† A 5000 V E.H.T. supply can be used, but the spark is much
less impressive and a Van de Graaff generator does better.

A spark provides ions If a spark passes through
an invisible jet of steam, it provides ions to
produce immediate condensation.

Procedure

(*i*) Show the jet without any spark. Use the
compact light source to form a shadow of the jet on
a distant wall or screen. Place the lamp 1 metre or
more from the jet.

(*ii*) Arrange two wires, from the Van de Graaff's
collecting ball and earth, to make a small spark gap
about 5 mm wide and about 3 mm above the glass
jet.

Start the steam supply and run the Van de
Graaff. Pupils watch the shadow.

Demonstration 86b
Ions start a cloud

Note The boiling flask must be far away from the
jet or the flame will make enough ions near to the
jet to spoil the demonstration completely.

Apparatus

1 500 cm³ flask with bung and glass tube	item 540
1 glass tube drawn to a jet (about 2 mm)	
1 Bunsen burner	
1 tripod	508
1 compact light source	511
1 L.T. variable voltage supply	21
1 retort stand, boss, and clamp	503–506

Cloud chambers Even at the beginning of this
century a few stout-hearted critics maintained that
the existence of atoms was not *proved*. It was
cloud-chamber pictures that convinced them that
there *are* atoms. The forks showed collisions of
individual particles.

And the many straight tracks of alpha particles
are the best evidence for our pupils that atoms are
mainly hollow, with a small massive nucleus.
Pupils should have had glimpses of cloud chambers in action before now, but they may not have
seen much of them, and there may be others who
missed them in earlier science. So all should now
watch tracks for themselves in diffusion cloud

chambers. But first they should see tracks in an expansion cloud chamber which will serve to show them what to look for.

As soon as the cloud chamber is shown, remind pupils to examine the pictures in their *Text*. These pictures now serve as prime evidence of the properties and behaviour of the high energy particles hurled out by radioactive nuclei when they disintegrate.

Demonstration 86c
Expansion cloud chamber

Apparatus

1 expansion cloud chamber	item 18
alpha particle source for cloud chamber	

Procedure

Since there are several forms of this apparatus which differ in operation, the instructions provided by the manufacturer should be followed.

Some forms require special alcohol, some operate with methylated spirit, some operate with water. The water-operated ones are preferable: they show clear thick tracks.

In using any cloud chamber it is essential to clear out 'old ions' – that is, ions left by alpha particles which went through the chamber long before expansion. Otherwise vague clouds will form on the old ions which have moved away from any well defined track. Apply a large voltage between the top of the chamber and the bottom or liquid surface. In some forms 200 V will suffice; in others, it is better to use electrostatic charging by rubbing the top window of plastic.

Class Experiment 87
Diffusion cloud chamber

Apparatus

8 diffusion cloud chambers	item 28
8 lamps	47
1 CO_2 cylinder	19/1
1 dry ice attachment	19/2
8 transformers	27

Procedure

There should be one cloud chamber for every four pupils. It is very important that they should have plenty of time for this experiment: it is something to be enjoyed and not hurried.

Pupils follow these instructions.

*　　*　　*　　*　　*

To set up the chamber, put methylated spirit on the padding inside the top of the chamber, using an eye-dropper. Also put a drop or two on the black base of the whole apparatus.

Ask your teacher to put a little dry ice from the CO_2 cylinder in contact with the plate. Screw the base cap on again and invert the chamber. The cloud chamber must be level. Place it on the three wedges provided, and adjust them to get it level. (If it is not level, you may see air currents moving in the chamber which you can use as guides in levelling.)

Put the top back on the chamber and give it an electric charge by rubbing it with a handkerchief or a finger nail. That will charge it sufficiently to make a strong enough electric field inside to sweep away old ions quickly and drive some ions of a new track into the sensitive layer.

Illumination is important. Adjust the lamps to illuminate a layer a few millimetres above the base plate.

Soon, usually within 30 seconds of setting up the cloud chamber, you will see alpha particle tracks coming from the weak radioactive source in the side of the chamber.

If the tracks are not sharp, rub the top again to improve the electric field.

*　　*　　*　　*　　*

Notes Surprisingly little dry ice is needed. Practice will show how much is required, usually about 5 or 6 cm^3.

To obtain solid CO_2 from the cylinder, fold a piece of closely woven cloth (preferably of dark colour) in the form of a bag. Hold this bag tightly round the nozzle of the cylinder and open the valve at full blast for 5 to 10 seconds.

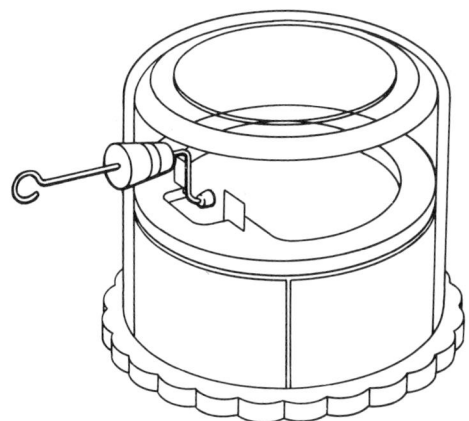

If the cylinder is of the siphon type, keep it upright. If it is of the type supplied for other uses, without a siphon tube inside, turn it upside down before opening the valve.

In schools where several classes are following the Nuffield programme, it may not be feasible to manufacture enough solid CO_2 from the cylinders. It will be necessary to order a block from the suppliers. Such blocks are obtainable, delivered by rail. See Appendix 5 of *Teachers' Guide 4* for addresses.

WHAT ARE ALPHA, BETA, GAMMA RADIATIONS?

Pictures *Pupils' Text* has photos of tracks of alpha particles, beta particles, and electrons ejected by gamma rays. The many straight tracks of alpha particles offer almost direct evidence for 'hollow' atoms in the air. The selected examples of tracks that made a fork – in a nuclear collision – deserve to be examined with the interpretation in the note under the picture.

There is also a photo of a nuclear event started by an invisible neutron. This is intended to raise interesting questions, in preparation for the discussion of neutrons later in this chapter.

Magnetic deflections Tell pupils that we can apply a magnetic field across the track of a beta particle in a cloud chamber and bend the path into a circular arc, as pupils did in measuring e/m for electrons.

But massive, energetic, alpha particles would need a magnetic field several thousands of times stronger to bend their path to the same curve as the 200-volt electrons in our lab. So *Pupils' Text* can only offer a special photo.

Identifying beta particles Pupils should see a qualitative demonstration of magnetic deflection.

Demonstration 88
Bending beta particles' tracks

Apparatus

1 scaler	item 130/1
1 thin window GM tube for beta particles	130/5
1 holder for GM tube	130/3
1 beta source	195/2
2 magnadur (slab) magnets	92B
1 iron yoke	92I
1 retort stand, boss, and clamp	503–506
1 lead block†	1M
1 cork	

†The lead block from the materials kit (item 1M) is just adequate. A block of larger area is preferable. A thick block of any other metal, or of glass, will suffice.

Procedure

Connect the GM tube to the scaler and hold it in a retort stand, as in the sketch, on one side of the lead block.

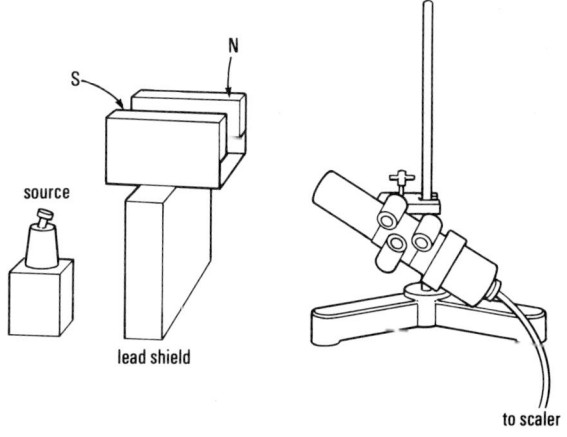

source

lead shield

to scaler

Place the beta source on the other side of the block. Fix it in a cork, pointing upwards as in the sketch.

The block shields the GM tube from direct radiation from the source. Only a low count rate will be observed.

Arrange a horseshoe magnet of two slab magnets on an iron yoke. Bring up the magnet above the lead block. Place it so that its field is horizontal, across the path of the beta particles. The catapult force of the magnetic field will make the beta particles follow a curved path. If the magnet is the right way round, many beta particles will now reach the GM tube; and the count rate will increase. If the magnet is the other way round the count rate will be as small as ever, due to a few scattered particles.

Turn the magnet each way to show the difference.

(It is interesting to physicists to determine which is the north pole of the U-magnet, thus settle the direction of its field, and then to use a rule for the catapult field to prove that beta particles carry a negative charge. But our pupils are beginners: all but a sophisticated enthusiast would get lost half-way through the argument – and that would spoil the experiment. We urge teachers to avoid that, though they might encourage the enthusiast to think it out on his own.)

Measuring charges and speeds Since the 'gun voltage' is unknown – built into the firing nucleus – we catch a lot of particles, count them as they arrive, measure the total electric charge caught, and then calculate the charge on one.

Alpha particles have charge $+2e$ compared with $-e$, the charge on the electron. Their charge/mass ratio fits with

$$\frac{2e}{4 \text{ times the mass of H atom}}$$

so we conclude they are He^{2+}. The magnetic field also gives their various speeds, comparable with 0.1 of the speed of light.

Beta particles are fast electrons, with the usual charge, $-e$. They are flung out with speeds ranging up to 0.98 of the speed of light. At such high speeds, their mass, to a stationary observer, is about seven times the normal value. Pupils may

also see tracks of slow beta particles bent into tight circles by a strong magnetic field.

Gamma rays have no charge and only one speed c, like all photons.

Identifying alpha particles *Pupils' Text* gives a picture showing alpha-particle tracks in wet helium, with one making a nuclear collision. The photograph is a lucky choice: a case when the fork occurred facing the camera directly.

The angle of the fork should be compared with the angle in an *elastic* collision between two large visible equal masses, one of them originally at rest. Show this with two very long pendulums carrying equal steel balls, or with equal ring magnets coasting on carbon dioxide on a level sheet of glass. (See **Demonstrations 41 and 42** in *Teachers' Guide Year 4*.) The comparison suggests strongly that alpha particles have the same mass as helium nuclei. And since the two tracks of the fork in the cloud chamber picture look alike we feel sure that alpha particles *are* helium nuclei, moving at high speed.

Radioactive changes Radioactivity has three characteristics that seem strange to pupils:

(*i*) The explosive event is *abrupt*. It happens suddenly, at random.

(*ii*) The atom changes to an entirely different atom, with different chemical properties – one element changes into another.

(*iii*) Atoms which have not yet decayed show no signs of ageing: they do not grow weaker or still more unstable as time goes on. Knowing those facts we can, with hindsight, expect exponential decay: a constant half-life.

EXPONENTIAL DECAY

One of the most important characteristics of radioactivity is exponential decay.* Both the size of a stock of radioactive material and the rate at which it decays die down exponentially, falling to half value again and again in the same time – or falling to $\frac{1}{10}$ value in equally regular, longer intervals. This suggests that the chance of an atom

*Here, as in every exponential change that we meet in physics, the story cries out for the simplification afforded by calculus. Yet by describing instability and using the concept of half-life, and by showing a real example, we can give pupils at the present stage an understanding which seems essential in the present state of the world.

disintegrating is constant in time.

We are looking at a series of many chance events, all with a standard, unalterable chance – or at least it appeared unalterable until we were able to bombard nuclei with streams of particles having such high energy that they can interrupt the course of normal radioactive life by effecting more immediate changes.

So the rate at which we count disintegrations is proportional to the total number of *unchanged* radioactive atoms at the moment. It measures the present stockpile.

Demonstrating half-life Restrictions on the availability and handling of radioactive materials in schools do not aply to the 'natural' radioactive element uranium and its salts. Uranium itself decays with an extremely long half-life, emitting a low-energy alpha particle. Its 'daughter' has a half-life of about three weeks. In a classic experiment in the history of radioactivity that daughter element was separated out chemically and its exponential decay measured. That process would be too slow to be convincing here.

That daughter has a daughter in turn with a half-life of just over 1 minute, emitting a beta particle. The latter element, the 'granddaughter' of uranium, can be separated from a solution of a uranium salt by extracting with an organic solvent. The half-life is short but a counter needs only a very small quantity to give an appreciable counting rate. Allowing for the time taken for the chemical separation we shall still obtain a useful sample from a small quantity of ordinary uranium salt (where the granddaughter will be 'in equilibrium'). There will be enough short-life protactinium (granddaughter) to provide easy counting with a scaler.

Demonstration 89
Exponential decay of a radioactive element

This experiment shows the exponential decay of a granddaughter of uranium (protactinium-234) with a half-life of 1·2 minutes.

The chemistry of the experiment

The early members of the uranium-238 family are involved. We make a solution containing uranium and its daughter (thorium) and the short-lived granddaughter (protactinium). We place this acidified solution in a plastic bottle and add an organic liquid, which floats on top of the aqueous solution. When the bottle is shaken, the organic liquid collects up about 95% of the short-lived granddaughter and some uranium but none of the daughter. As the granddaughter decays, its atoms emit beta particles which travel through the plastic wall of the bottle, reach the GM tube, and are counted.

The counter does not detect the alpha particles from uranium or the low energy beta particles from the daughter element. It only records the betas from the granddaughter.

$$\overset{238}{\underset{92}{}}U \xrightarrow{\alpha} \overset{234}{\underset{90}{}}Th \xrightarrow{\beta} \overset{234}{\underset{91}{}}Pa \xrightarrow{\beta} \overset{234}{\underset{92}{}}U$$

Apparatus

1 scaler	item	130/1
1 holder for GM tube		130/3
1 thin window GM tube		130/5
1 stop-clock		507
1 retort stand, boss, and clamp		503–506

1 small polythene† bottle (30 to 50 cm^3 capacity)
uranyl nitrate (or uranium oxide dissolved in nitric acid)
concentrated hydrochloric acid
iso-butyl methyl ketone, or amyl acetate

† A polypropylene bottle is somewhat more resistant than polythene to attack by the liquids, so it would be preferable; but a polythene bottle can be used provided the liquids are not kept in them for more than a few weeks. Avoid polystyrene bottles: the liquid attacks them.

A cheap bottle has the advantage of thin walls which let more beta particles through with sufficient energy. But if its cap has a cork lining screw a small piece of thin polythene under the cap for protection.

Preparation

The plastic bottle should be nearly full, containing equal volumes of organic reagent and acidified uranyl nitrate solution.

For every 10 cm^3 of acidified solution, dissolve 1 gram of uranyl nitrate in 3 cm^3 of water and add 7 cm^3 of concentrated hydrochloric acid. Those proportions are only rough guides.

The mixing can be done in a beaker – it is quite unnecessary to use a special funnel. Pour the mixture into the plastic bottle and add an equal volume of iso-butyl methyl ketone or of amyl acetate. Screw the cap on the bottle tightly.

Procedure

Place the bottle in a tray lined with absorbent paper.

Support the GM tube by a clamp on the tube holder (*not* on the tube) and tilt it so that it points steeply downward, slanting towards the neck of the bottle.

Beta particles come out to the GM tube through the thin wall of the plastic bottle.

Shake the bottle vigorously for about 15 seconds and place it in position. As soon as the two layers of liquid have separated, start the scaler and

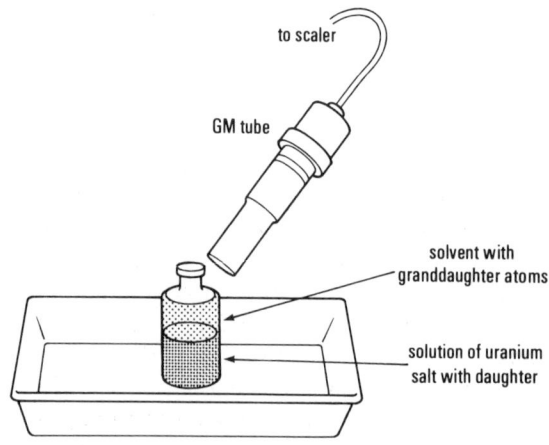

to scaler

GM tube

solvent with
granddaughter atoms

solution of uranium
salt with daughter

let it count for 10 seconds. Take counts at 10-second intervals without stopping, or take a 10-second count every half-minute. Record each count.

Also record the 'time of day' – that is, the total time from the start of counting to the start (or to the mid-time) of each count. That tells us, so to speak, the 'age' of the source.

There is no need to continue counting longer than 5 minutes, when the activity will have dropped to less than 10 per cent of its initial value, apart from background.

Allow for background radiation. Wait till 10 minutes from the start of counting (when the activity will be down to $\frac{1}{4}$ of 15 per cent of its initial value) and set the scaler to count for, say, 5 minutes, with the bottle still in position. That count will provide an average value of the background, some of which comes from the lower liquid.

Subtract the background (*reduced to a 10-second rate*) from each of the earlier counts. Discuss the obvious decay and try to find a rough value for half-life.

Eager sophisticates in the class might be encouraged to take common logs and plot a graph of LOG OF CORRECTED COUNT against TIME OF DAY at the mid-time of each count. This is likely to give a satisfying straight line.

The experiment can be repeated at once. In the 5 minutes of counting the protactinium in the aqueous layer. is replenished to 95% of its equilibrium value. Just shake the bottle again, and start counting.

The other half of the story, growth to equilibrium can be followed if the GM tube is placed *below* the bottle. Then counting with the scaler will show protactinium being replenished from its thorium parent. However, while a physicist will enjoy seeing that, beginners are likely to be confused and even lose the point of exponential decay – so we do not recommend it at this stage.

THE GREAT SCATTERING EXPERIMENT: NUCLEAR ATOM MODEL

Atomic investigators Radioactivity is indeed a rich field for learning about atoms. As well as the instability and transmutation of radioactive elements they also provided alpha particles to act as investigating missiles to show us the structure of ordinary atoms. Nowadays big machines, 'accelerators', provide still more energetic particles to probe the interior of atoms, but the nuclear atom model was developed to fit the surprising behaviour of alpha particles.

RUTHERFORD MODEL

Pupils have heard of electrons and positive ions, so that a 'plum-pudding' model of atoms may well seem reasonable; but if we proceed already to a description of a nuclear model without showing compelling evidence, we shall find ourselves in a hopeless field of assertion.

Evidence for a hollow-atom model One might say to pupils: 'Our first line of evidence is as we said above the almost direct one of cloud chamber

144

pictures. Looking at a cloud chamber in action with alpha particles you will see straight tracks and straight tracks and straight tracks, again and again, and never a massive nuclear collision. While an alpha particle makes 100 000 'minor' collisions, it almost never makes a '*major*' one in which it bounces away on a path in a new direction.

'In these minor collisions, the alpha particle tugs an electron off an air molecule as it hurtles by, and thus produces a positive ion and soon, a negative one; drops of water can condense on these ions and make the visible track.

'Since the track of the alpha particle is not noticeably bent by these collisions, we think it must have hit something of trivial mass each time – in fact an electron some 7000 times less massive than itself.

'However, if we take a great many photographs of cloud-chamber tracks we see occasional examples of a fork in which the alpha particle's path is deflected. The enormous number of straight tracks is evidence for a *hollow* atom, with detachable electrons somewhere in its outer region; both the rareness and the big angles of the forks are evidence for a very small *nucleus* in which most of the mass of the atom is concentrated.

'But that knowledge came after the great scattering experiment had forced Rutherford to devise his nuclear model.

'**The experiment** Rutherford encouraged two young assistants, Geiger and Marsden, to count the alpha particles scattered by a thin target of gold leaf. They fired a stream of alpha particles from a small gun at the gold leaf in a vacuum. Most of the particles went straight through – as with cigarette paper. But a few bounced off the direct track at a noticeable angle. (You know that from cloud-chamber pictures.) And, great surprise, a very small number bounced back from the gold. That was unexpected and it could not fit with the model of gold atoms as a lot of electrons embedded in a pudding of positive charge – there could not be a close enough encounter to swing the path of the missile round and back.

'So Rutherford was forced to imagine a new model. (That is how most great advances in theory are made: the scientist doesn't dream up a new model from nowhere: he is forced to make the change, forced by new experimental discoveries.)

'The new model is a nuclear atom, a tiny, positively charged core with most of the atom's

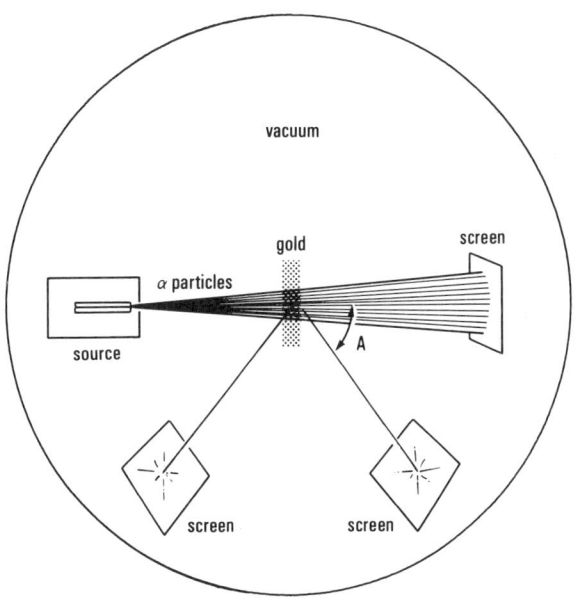

mass, and electrons far out from the core. Alpha particles that bounced back had swooped inside the electron groups and had been repelled by the core at close quarters. Why *repelled*? Because both the alpha particle and the core are, according to the new guess, positively charged. Two positive charges repel with an inverse-square law force. But would that same type of force still hold at tiny distances inside an atom? Rutherford saw that Geiger and Marsden's measurements could answer that question. He calculated how many alpha particles would be scattered through various large angles if an inverse-square law of repulsion controlled the orbits of the exploring alpha particles. He compared the measurements with his predictions.'

The evidence Pupils should look at the table of measurements (in *Pupils' Text*) and the comparison of those with theory. That result would have delighted Kepler. Like Kepler's Law III, the result vouches for an inverse-square law over a vast range.

Models of scattering Although the rare backward scattering was a surprise to Rutherford, he knew quite well what the bullets were that he was firing; and he could see how the results would force him to make a new picture or model for atoms. But our pupils approach this evidence with no such preparation. We need to help them with illustrations.

145

Ask them whether one could find out if cannon balls are concealed in a truss of hay, by firing rifle bullets at the hay. Ask them to think of rolling marbles along a slightly sloping table with a few spikes projecting upwards, as in a pin-ball machine. What would happen to the marbles?

Show them Rutherford's teaching model with magnets. Also show a model – nearer to the real event – with electrostatic forces.

Rutherford's magnet model In his university lectures, Rutherford illustrated his own work on alpha-particle scattering with a giant model which he showed with great delight. Magnet poles represented the charged nucleus of a gold atom and the charged alpha particle flying past. We should show pupils a model like that.

{The ideal, but impossible, arrangement would be a very long pendulum with its bob an isolated North pole hung so that it swings just above a stationary isolated North pole. The inverse-square repulsion between the two North poles would make the bob follow a hyperbolic path, except for the effect of gravity on the bob. With a very long pendulum the restoring force due to gravity might be so small compared with the magnetic repulsion that the orbit closely matched an alpha particle's hyperbola.}

{However, with real magnets there will always be a pair of poles, or the equivalent, and if the magnets are short the repulsive force between the pendulum magnet and the fixed magnet will be nearer to an inverse fourth-power force; then the restoring force due to gravity – which increases with the displacement – will dominate the motion except at very small approaches.}

{Therefore in a model with real magnets it is important to make each magnet long (so that its south pole is less important) and strong. For strength it would be best to use electromagnets as Rutherford did in his own model. One could easily arrange this for the stationary magnet by using the I-yoke of the demountable transformer with a coil fed by d.c.; or, better still, a tall bar of soft iron with a coil on it. Hanging another electromagnet on the pendulum is not so easy and the trailing leads may seem confusing. A simpler form with permanent magnets is easier to set up, and still well worth showing.}

Demonstration 90X
Rutherford's magnet model of the path of an alpha particle inside an atom

For the pendulum, use a very long light rod or tube and attach at the lower end either a long bar magnet or a chain of shorter ones joined head to tail. For the stationary magnet, again use a long bar magnet if one is available, or make up the equivalent of short magnets. There, large mass does not matter, so a bundle of magnets side by side, all with their North poles upward, will be advantageous – or use electromagnets.

Since the choice of magnets will vary according to the resources of the lab, the list here merely says magnets, and the diagram shows an assembly of short bar magnets.

Apparatus

3 retort stand rods	item 504
2 retort stand bases	503
2 bosses	505
1 clamp	506
1 G-clamp	44/1
magnets	
1 light rod†	
thin walled rubber tubing	

† The pendulum rod should be at least 1 metre long, either wood or aluminium tube, about 1 cm diameter.

Preparation

Suspend the pendulum rod by a short loop of thread (a few cm) from a horizontal rod on a retort stand. Clamp the base of the retort stand to the table.

Attach a bar magnet, or an equivalent assembly, to the lower end of the pendulum rod with rubber tubing or with tape.

Arrange the magnet, or equivalent group, so that its North pole is upward and only slightly below the North pole of the pendulum's magnet.

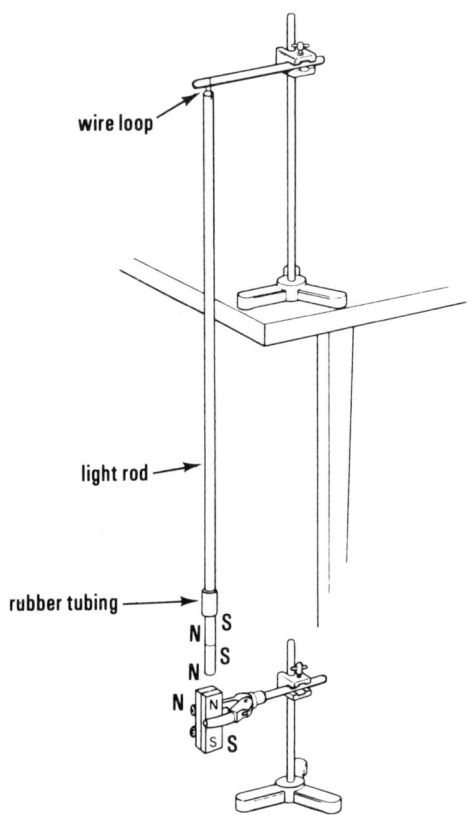

wire loop

light rod

rubber tubing

N S
N S
N S
N S

Procedure

Draw the pendulum aside and release it. Show several patterns of orbit. It is probably best to start with a collision that is almost head-on and then show wider and wider collisions with smaller deflections.

To minimize gravity control, limit the pendulum to fairly small amplitudes, but start it with a push. Then the initial and final parts of the path will look almost straight, with the bend showing only during close approach – as with the alpha particles.

Demonstration 90Y
Electrostatic model of alpha particle
scattering

Apparatus

1 table-tennis ball coated with Aquadag	item 131E
1 Van de Graaff generator	60/1
fine nylon thread	51E

Preparation

Make sure the table tennis ball is well coated with Aquadag to make it conducting. Attach a long nylon thread to it with Sellotape. Suspend the ball by the thread, preferably from the ceiling so that the suspension is as long as possible.

Set up the Van de Graaff generator with its sphere level with the suspended ball and quite close to it.

Procedure

Explain that the suspended ball represents an alpha particle, repelled by the Van de Graaff sphere when both have charges of the same sign.

Run the Van de Graaff; let the small ball touch the sphere so that it becomes charged.

Pull the small ball out to one side with an insulating rod and release it. Pupils watch the orbits made by the small ball.

(With a large class who would see the orbits foreshortened, it may be better to arrange a compact light source and a mirror at 45° to cast a shadow on the ceiling.)

Film 91
'The Rutherford model of the atom'
(*OPTIONAL*)

The Nuffield Advanced Physics project collaborated with Rank and Mullard to produce this film, in which the experiment by Geiger and Marsden is repeated. The film uses school apparatus as far as possible, though the source is far too strong for school use. Measurements are taken at several angles, and the film shows the variation of number of particles caught with angle and compares that record with the number predicted for an inverse-square law.

This film offers very good teaching; but we list it as optional because schools may not consider the expense justifiable for O-Level.

Commentary on the scattering experiment
We should not let pupils think that this is just a

routine measurement in atomic physics; we should describe the experiment, show the results, and even explain the general idea of the theory, so that they see it as one of the great turning-points in physics. It changed our picture of atoms permanently. To do it justice in our teaching we must show pupils some of the real story and not just make assertions – otherwise we shall seem to be telling fairy stories.

On the theoretical side, we merely say that Rutherford assumed an inverse-square-law repulsion between the big electric charge on the massive nucleus of the gold atom, and the smaller electric charge on the alpha particle flying past it. That is equivalent to Newton's assumption of an inverse-square attraction between the massive Sun and a planet. But instead of the simple circular orbits which serve approximately for planets, the change to a repulsive force predicts a different shape: hyperbolas. The alpha particle sails in, bends round a corner and sails out again on another almost straight track in a new direction. The simple calculation with circular orbits that predicts Kepler's Law III becomes complicated.

Furthermore instead of a few individual planets, each with measured orbit and period, Rutherford had to use hordes of little alpha particles to give him a statistical test. He made his theory predict the number of particles that an observer would count on a receiving screen in various directions, in some standard time. In calculating that prediction he simply used an inverse-square law of repulsive force and Newton's laws of motion.

We ask pupils to take our word for this and we join them in comparing theory and experiment by looking at the table in *Pupils' Text*.

Remind pupils that this was *not* an investigation directly concerned with radioactivity. Alpha particles were being used as explorers to test the inverse-square-law field over a great range of distances inside an atom.

When, instead of alpha particles, we use protons that have been accelerated by a modern machine, such as a cyclotron, we have a much bigger choice of energies for the bombarding particles, and the inverse-square law seems to break down at close approaches.

The nuclear charge The scattering of alpha particles not only provides us with clear evidence for a nuclear atom, but enables us to measure the nuclear charge. Chadwick used thin sheets of copper, silver, and platinum, instead of gold, and measured the scattering of alpha particles by each. From his counts he calculated, with the help of Rutherford's theory, the charge on the nucleus of each of those three kinds of scattering nucleus. His results were:

copper 29·3 electron charges,
silver 46·3 electron charges,
platinum 77·4 electron charges,
with expected errors about 1 per cent.

The serial numbers of those elements, arranged in order of atomic masses are: 29, 47, 78.

Nowadays, we *define* the atomic number as the nuclear charge (measured in electron charges), but originally it was merely the serial number of the element. So Chadwick's measurements showed that the nuclear charge *is* the serial number, or the atomic number, now called Z.

Another measurement of atomic number Soon after Rutherford's announcement of his atomic model, when Bohr was describing the early form of his model, Moseley made measurements, with crystals, of wavelengths of characteristic X-ray lines of several elements. Interpreting them on early Bohr theory he obtained numbers for nuclear charges which supported Rutherford's suggestion.

Believing in models Rutherford's model was both suggested and supported by experimental knowledge – although it presented a paradoxical problem over stability.

In this matter of making models, it is important to help pupils to see that the models and their charges have been suggested in part by experimental knowledge. Then pupils should feel some confidence in models; yet they should respect our warning that models may change.

CHANGES FROM ELEMENT TO ELEMENT

Describe, very briefly, the changes when an unstable element emits a nuclear particle.

Even at A-Level, we should not let this expand into a new form of rote-memory science in which nuclear reactions are learned by heart. In selecting examples to illustrate various types of change, we should survey the great variety known today and choose examples that seem simplest rather than adhere to the original cases of the historical development.

Transmutation We, as physicists, take radioactive changes from one element to another in our stride, but we should present them to pupils as an amazing break with the whole tradition of nineteenth-century chemistry, a change of the 'ultimate' atoms, a transmutation of immutable elements. We should emphasize the complete contrast of chemical properties between parent and daughter elements.

Natural transmutation: radium's daughter element The alpha particle which emerges from a radium atom comes out with such huge energy that we are sure it came from the nucleus. And that is confirmed when we find that the 'daughter' atoms – the atoms that were radium until they ejected an alpha particle – are entirely different chemically. They form a dense, inert gas in strong contrast with the properties of radium as a heavy metal.

Placing radium in the calcium-barium family of the Periodic Table, we find we must move two columns downhill to place the product, radon gas, in the column for helium, neon, etc. An alpha particle carrying away charge $+2e$ would produce just that shift if the serial numbers of elements in the chemical series *are* the nuclear charge numbers.

Again, radium has long been plentiful enough for good determinations of its atomic mass number 226; and the density of radon gas has been measured and found to agree well with the predicted 222. The story of this and all the other radioactive changes agrees with our view that there is transmutation from element to element.

When we have a mixture of parent element and daughter element, which have different chemical properties, we can separate them by ordinary chemical methods. We seldom have handled, or wish to handle, large enough quantities for visible chemistry, so we add stable, non-radioactive material of the same atomic number as one of those elements – a common inert isotope – and then we have enough for ordinary chemistry. Yet we can trace the fate of the radioactive isotope with a Geiger counter. And we find that the radioactive material moves everywhere with the inert material of the same element – the chemistry of the atoms' outer electronic regions is essentially identical.

On the other hand, when we use chemistry to separate some pure parent stock from a daughter element, we find, if we wait some time, the daughter element appearing among what was pure parent stock. In other words, we see the result of continuing transmutation.

This is not something that we can demonstrate directly to pupils because it means handling radioactive materials, mixing them with other materials, carrying out chemical separations, and still making sure that nothing radioactive gets thrown down a drain or used in any dangerous manner. We can only *tell* our pupils about radioactive changes and chemistry.

Manufacturing unstable isotopes: tracers Sixty years ago, radioactivity was known as a peculiarity of a few heavy elements – the last dozen at the end of the chemical series are unstable. Now that we can bombard samples with high speed, high energy, protons or neutrons (provided directly or indirectly by an accelerator such as a cyclotron or a linear accelerator) we can make unstable isotopes of every element in the whole chemical series.

This has opened up a tremendous new field of 'nuclear chemistry'. Though it is fascinating and useful, we do not suggest that this should turn into a new field for 'learning things by heart' in our science teaching. But pupils should learn why radioactive isotopes that we now make are so useful: they serve as tracers, like tags or luggage labels that enable us to follow any element we wish through processes such as digestion and circulation in a living body or in commercial manufacture.

If we add a small quantity of radioactive sodium to a large sample of common sodium in salt, we can then follow with a Geiger counter the progress of sodium through any system, even a human body.

Tracers may also assist diagnosis. For example, it is often difficult to tell when and where internal bleeding (haemorrhage) occurs in a patient. Atoms of chromium are readily taken up by the red cells of the blood. So the radioactive isotope chromium 51 can be used to label the patient's blood. If haemorrhage is occurring, the radioactivity will increase in that region of the body and the rate at which it increases indicates the volume of the blood being lost.

In another example, barium 137 (with a half-life of 127 days) is injected into a patient's vein and immediately enters the heart. A counter directed at the heart detects the tracer and enables the working action of the heart to be examined as the count rate rises and falls with the heart beat.

LATER TOPICS IN *PUPILS' TEXT* CHAPTER 10

Pupils' Text describes some later developments in the field, including the discovery and behaviour of neutrons and, in particular, the relationship between neutrons and fission, the application of this discovery to the production of electrical energy in power stations, the production of plutonium and the uses and dangers of that element, and the hope for fusion, together with some comments on the risks and the benefits of these applications.

CHAPTER 11
Waves and particles

In earlier chapters we have examined the wave behaviour of light and the particle behaviour of entities such as the electron and the alpha particle. We need to be careful to stress that light *behaves* as a wave, that electrons *behave* as particles, and to avoid direct statements of the form 'light is a wave' for, in the sciences, we now realize that our concern is with answering questions in the form 'How does this or that behave in a given situation?'.

Before ending the course it is necessary that we should show our pupils that light also behaves as particles do, and that material particles may themselves demonstrate wave behaviour. Then we shall have provided our pupils with an essential introduction to a modern view.

For our present teaching the photoelectric effect gives much the clearest evidence for 'packets of energy', quanta, of light. Nevertheless, at this stage in their learning, pupils must expect to rely to a greater extent than hitherto on vicarious evidence; they must accept experimental evidence gathered by other scientists and published to the world.

EARLY EVIDENCE AND ORIGIN OF THE QUANTUM THEORY

Several phenomena pointed towards a strange restriction on interchanges between radiation and atoms, in the early part of this century.

The idea that radiation energy (or at least its interchanges with energy of matter) comes in packages of amount proportional to frequency first arose in Planck's mind and he used it successfully in fitting a theoretical prediction to the experimental curve for the distribution of energy in the spectrum of a perfect radiator. Classical theory completely failed to predict the experimental curves until the quantum restriction was imposed in addition. This is far too difficult an avenue into quantum theory to carry any conviction with young pupils.

Then specific heats, both of solid elements and of gases, showed strange changes with temperature which were not predicted by classical mechanics (equipartition) but could be accounted for successfully by imposing the quantum rule on rotational motion and vibrations of molecules.

Then the photoelectric effect pointed in the same direction – or rather, was found to be pointing, when Einstein applied his clear vision to it.

Series in spectral lines, measured and decoded, waited for some explanation, and they too fitted into the quantum scheme when Bohr thought out his model for atoms. X-rays and radioactivity added signposts pointing to quanta too.

It is clear now that *all* electromagnetic radiation carries its energy in quanta of size $h \times$ frequency; and that periodic motions of molecules (spins and vibrations) also have their energy in one or more quanta. The quantum constant is another of the atomic constants in the universe.

THE PHOTOELECTRIC EFFECT: WAVES INTERACT WITH MATTER

We now turn to a phenomenon that is very fruitful and easier for pupils to understand. They have seen photocells at work – in applications where light releases a horde of electrons from a sensitive surface in a vacuum, and the horde acts as a current which we can set to work. That might be called the 'wholesale' photoelectric effect.

Next we consider in detail the 'retail' effect, and pupils hear of light flicking electrons out of a metal, ultraviolet light tearing them out with the crack of a whip, X-rays hurling them out. This strange interchange between radiation and electrons throws much light on the microphysical world.

Demonstration 92
Waves interacting with matter; the wholesale photoelectric effect

Apparatus

1 gold-leaf electroscope   item 51A
1 zinc plate attachment   190
1 wire mesh
1 E.H.T. power supply   14
1 piece of fine emery cloth
1 sheet of glass   58F
1 ultraviolet lamp   189

An alternative source of ultraviolet light is provided by the small arc described in *Teachers' Guide Year 4*, page 230.

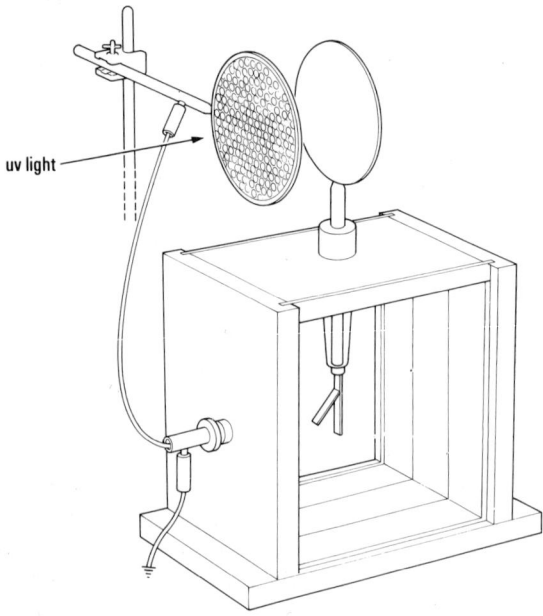

uv light

Procedure

Set up the gold-leaf electroscope so that the pupils have a clear view of the leaf (This may be shadowed onto a ground-glass window in the electroscope using a 12-V 24-W lamp.)

Thoroughly clean the zinc plate with the emery cloth and attach it to the electroscope. Support the wire mesh a few inches away from the zinc plate. This mesh is connected to the case of the electroscope which is earthed.

Charge the plate on the electroscope negatively. Then illuminate the plate with the ultraviolet lamp. Observe the effect on the gold leaf.

Repeat the experiment with the plate and electroscope charged positively.

Finally repeat the experiment with the plate negatively charged again, but when the charge is clearly seen to be leaking away, interpose a sheet of glass between the light source and the charged plate.

This experiment suggests some of the photoelectric effect story, but it does *not* show that the negative electricity is coming out in particles (electrons); it does *not* show that the light is arriving in bundles of energy (quanta). It only suggests that there is some connection between the wavelength of the light and its efficacy in ejecting negative charge.

A cloud-chamber photograph included in the *Pupils' Text* shows the track resulting when X-rays pass through a gas. The radiation is monochromatic Cu K which releases 5 and 8 keV electrons from the argon, giving tracks less than 1 mm long. Tracks outside the X-ray pencil are due to other causes. The photograph, which is by E. J. Williams, is helpful on two counts: it shows the tracks left by the photoelectrons and also how the beam of X-rays is absorbed.

It would be good if we could show protons arriving one by one and that, for light of a given wavelength, the maximum energy of the photoelectrons is the same whatever the intensity of the light. Teachers may wish to experiment with a Geiger tube or a solid-state detector as a detector of ultraviolet photons. Where a laboratory has a Geiger tube with a very thin end window it is possible to detect such protons from a carbon arc, or from a flaring, lighted match, or from the Sun itself. The normal radioactive background count cannot be avoided so that the story is far from clear. Alternatively a solid-state detector can be tried. If fitted with a discriminator, this should be adjusted as for the detection of alpha particles.

Such an experiment can do no more than suggest that the detector responds to pulses of radiation which can cause ionization and that these pulses appear to arrive randomly.

In the absence of an experiment, it will be necessary to introduce the evidence gathered from the experiments of other people – a standard procedure in scientific work. In the first edition of the Nuffield programme, reliance was placed on two excellent P.S.S.C. films, *The photoelectric Effect* and *Photons*. However, experience has shown that the hiring of these films does present schools with difficulty and we now consider them as optional luxury aids.

We do not recommend here the series of measurements with special photocells which is commonly used in more advanced teaching; the experiments do not offer a sufficiently clear-cut story because of the contact p.d. which appears; and there are other difficulties.

Among experiments which do provide some assistance we suggest the following two.

Demonstration 93a
Counting gamma-ray photons

Apparatus

1 scaler	item 130/1
1 GM tube holder with gamma tube	130/3 and 6
1 gamma source	195/1

Note It is helpful if the arrival of the photons can be made audible through a loudspeaker so that pupils can listen to the effect.

Procedure

Connect the GM tube to the scaler. Place the gamma source 20 cm from the tube and note the count rate. Increase the distance between the source and the tube to, say, 40 cm and again note the count rate. Longer distances should also be tried.

Demonstration 93b
Photons and photochemistry

Apparatus

1 piece of each of the following Cinemoid colour filters	item 205

Primary red (No. 6)
Primary green (No. 39)
Primary blue (No. 20)
Bromide paper and facilities for processing

Procedure

Cut pieces of the colour filters about 10 cm by 5 cm and arrange them side by side on a sheet of bromide paper in the dark room. Expose the paper through the filters to white light for a few seconds. Develop and fix the paper. The final print will reveal that the red filter shields the sensitive paper from any photochemical effect.

Detailed knowledge Our present picture of the photoelectric effect emerged slowly from a variety of observations and suggestions.

We may quote Einstein's acceptance (1905) of a proposal of Max Planck's that the energy in electromagnetic radiation comes in packets of a definite size (quanta) which, he said 'penetrate into the surface layer of the body and their energy is transformed, at least in part, into kinetic energy of electrons. The simplest way to imagine this is that a light quantum delivers its entire energy to a single electron.' The sizes of these quanta were, he suggested, related to the wavelength of the light by the equation:

energy of the photon = constant $(h) \times$ frequency

where h is Planck's constant.

To our pupils each aspect of the phenomenon is strange and new and needs a clear statement from us that it *is* a new piece of knowledge. We should teach our pupils that in the retail photo-electric effect:

1. The particles ejected are electrons, with the usual value of e/m once again a universal ingredient.

2. For a given illumination, the electrons emerge with a variety of speeds, the slower ones having probably lost energy in collisions as they travel through outer layers of the metal.

3. The maximum speed of electrons is determined by the wavelength of light used and not by the intensity. Brighter light only produces *more* electrons – to everyone's surprise, in the early days – and not *faster* electrons.

4. The maximum energy of ejected electrons appears, after an allowance has been made for energy to escape, to be proportional to the frequency of the light. This is the basis of Einstein's equation.

5. When the light is first turned on, there is no delay in production of electrons, as one would expect if a continuous stream of light waves had to build up enough energy in the metal to eject each electron in turn.

Experiments with very weak light tell an impressive story. Sometimes an electron is ejected very early, almost at once when the light is switched on; sometimes no electron emerges till very late, after the weak light has been shining for some time; in general there is a random distribution of timing. If weak light is turned on and then off after so short a time that we could not expect its total energy to eject a single electron, we still see an electron ejected, during the illumination, *sometimes*. We are forced to picture the energy of the light arriving in small quanta with random spacing in time.

All these are aspects of the photoelectric effect which are new and strange to pupils. And we must rely on descriptions.

Pupils will be very familiar with the idea that the apparent smoothness of ordinary matter arises from the minute size of the atoms of which it is made up and from the vast numbers of such atoms. Now they begin to recognize that the seemingly smooth flow of events arises from minute erratic changes at the sub-microscopic level. Matter and action are alike in that all the regularity of materials and of the flow of events results from the adding together of vast numbers of elementary units.

X-rays; particles interact with matter We do *not* advise schools to buy X-ray equipment for use at this level. It is expensive and hedged about with safeguards. Pupils will know that X-rays can cast shadows of bones in flesh, and that the rays can blacken photographic film. Yet we need to give them a short description of the production of X-rays: electrons from a hot filament accelerated by a large p.d. gain huge kinetic energy; on reaching the target they lose their K.E., nearly always converting it to heat in the target. Just a few electrons, far less than 1% of the stream, convert their K.E. to a quantum of radiation as they come to rest.

X-rays and crystals Pupils will be familiar with the use of diffraction gratings to produce spectra of visible light, and will have seen how a piece of fine cloth, used as a two-dimensional grating, can produce a pattern of spots. Mention has been made

in Chapter 9 of the somewhat similar patterns caused when X-rays are reflected (diffracted) from the planes of atoms in a crystal.

The pattern from a single crystal (von Laue pattern) is so different from the rings given by a polycrystalline material that pupils will need help to appreciate the transition.

Class experiment and demonstration 94
Patterns from gratings

Apparatus

16 coarse gratings	item 191/1
1 compact light source	21
1 12-V supply	27
1 rotating grating	198
1 slide projector	
1 screen	

The slide carrier is provided with an improvised dark slide of card with a 3 to 5 mm circular hole at the centre.

Procedure

The compact light source is set up at one end of the room so that all may see the filament. The pupils are then asked to view the filament through a grating. They then turn the grating through a few degrees; then a few more ... They are asked to speculate on the pattern which they would see if the grating were to turn continuously.

We then ask what they think they would see if they looked at the distant source through a grating of woven cloth and kept that turning round. We use a device for that which consists of a small ball-bearing assembly, with a short piece of metal tube pushed into the inner bearing cylinder to act as handle. A small piece of woven cloth or a piece of the coarse diffraction grating material is placed across the outer cylinder of the bearing. As the pupil looks through the tube at the filament, he spins the outer cylinder so that he sees the light through a revolving grating.

Alternatively, focus the projector lens so that there is a sharp image of the circular aperture on the screen. Hold the rotating grating close to the projector lens and observe the pattern on the screen. Then rotate the grating in its holder and observe the pattern again.

If a sheet of plastic grating replica can be obtained, a better demonstration can be offered. Cut a piece of the grating replica into small pieces (a few mm across) of many shapes. Stick these jigsaw fashion on a sheet of glass (a 5 cm glass slide to which double-sided Sellotape has been stuck is excellent) to provide a higgledy-piggledy pattern of small gratings. Complete the sandwich with a second slide, so ensuring that the grating assembly is flat. Place the composite grating in the beam of the projector and view the pattern on the screen.

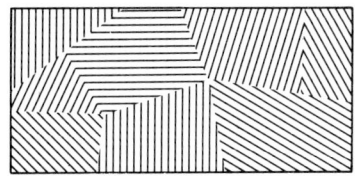

Pupils are then asked to examine photographs on pp. 181–2 in the *Pupils' Text*. The first shows the diffraction pattern obtained when X-rays fall on a single crystal of potassium alum: the second shows the pattern obtained when X-rays fall on a layer of powdered potassium alum. The two patterns, which were photographed with CuK alpha radiation (exposure time 1 hour), are evidence for the diffraction of the X-rays at the planes of atoms within the crystals involved.

The next photograph to be considered shows the screen at the end of an electron tube in which a beam of electrons is falling on an assembly of tiny graphite crystals. It was obtained by photographing the Teltron electron diffraction tube in action and where a school possesses such a tube for sixth-form use it may well be demonstrated here. (For details see *Nuffield Advanced Physics*, Unit 10.)

The pattern resulting from the passing of a beam of electrons through the carbon is very similar to that obtained when a beam of X-rays passes through a powdered crystal. It suggests that electrons can suffer diffraction as well as X-rays and other waves.

Electrons appear to show wave behaviour.

A further pair of photographs compares the interference pattern of light in a Young's double slit experiment with the interference pattern obtained in an experiment with electrons from a double slit source by C. Jönsson.

We know that electrons carry energy – and, we suppose, momentum – when they are moving. Yet these moving electrons seem to be guided to an interference pattern just like waves of light; or, rather, just like photons of light.

In the microphysical world, at the level of atoms and parts of atoms, the wave behaviour of a moving particle is very important; wave patterns guide the particles to produce diffraction and interference patterns.

In the macrophysical world, large particles (cricket balls, men, aircraft) do not display wave behaviour; wavelengths associated with such particles at usual speeds are so extremely minute that we cannot expect to observe the diffraction or interference patterns associated with them. Nevertheless we must now accept that the behaviours of waves and particles in the microphysical world are not entirely separate. Moving particles *do* follow wave directions and it is the wave which tells us where to find the particle. The particles are 'guided' by matter waves.

Such a description of matter will seem strange and perhaps without point to our pupils. Nevertheless, it is so much a part of modern thought – not solely of physics – that we feel it right to provide a simple introduction.

In discussions with pupils we should do well to remember the complementarity principle of Neils Bohr, who pointed out that ideas of waves and particles complement one another. By this he meant that the description of observable phenomena requires that these apparently irreconcilable concepts be applied in turn, since the descriptions they offer are never simultaneously applicable to the whole. Many observations in atomic physics can be treated using the particle model on its own. Others require the wave model. Both models prove to be useful and, despite their contradictory nature, must be used alternately. They are never in direct conflict because it is impossible to determine at the same time all the information required to make the two images precise (Uncertainty Principle). The more precise we wish to make one picture through observation, the less precise the other becomes.

Define the wavelength of an electron sharply enough and the attempt to apply the particle model will surely fail. Localize that electron definitely enough and the wave model fails. To put it another way, ask a wave question and you will get a wave's reply. Ask a particle question and you will get a particle's reply.

The two aspects of the electron both contradict and complement one another: both aspects are needed for a complete description.

ATOM MODELS

Bohr model Not long after Rutherford set forth his nuclear model for atoms, Niels Bohr made a brilliant move to bring that model into accord with some experimental knowledge of quanta, and he produced a very fruitful theory.

We should not try to describe the Bohr model in detail to our pupils, still less should we try to carry them through any calculations. For one thing it requires too large a background of new knowledge; and for another thing it is out of date. It does not agree with our present view, though it has played an essential part in leading physicists towards it.

Wave patterns in atom models We turn from a running electron wave – which experiments forced us to imagine – to the idea of stationary electron waves in an atom. That is how the new idea of guiding waves carried the Bohr atom model towards a great modern development.

In writing down rules that seemed *necessary* for a Rutherford atom model, Bohr had to state several novel rules, such as the requirement that an electron has a choice of definite, stable orbits, and the rule that in switching from one orbit to another the electron emits the difference in energy as a single quantum of amount (constant, h) $\times$ (frequency). The latter rule simply insisted on the quantum relation that was already known. But the definition of stable orbits seemed rather arbitrary in Bohr's hands. There must be some such restriction – line spectra tell us that – but could one give a sensible rule for defining the particular orbits? Bohr did that with a rule for 'quantizing' angular momentum, which was at best plausible.

With the concept of guiding waves for moving electrons, it was easy to give a much more appealing rule for choosing orbits: the circumference of the orbit must contain a whole number of wavelengths, so that the electron's wave pattern is a stationary wave. That 'explained' why the Bohr orbits were stable, and told us we could not expect to locate an electron precisely at a point on a sharp orbit; and it led to some of Bohr's predictions of orbit sizes and energies. But when a fuller wave model was developed it no longer agreed with Bohr's model; but it has proved fruitful in successful predictions or explanations. So the first, ingenious use of stationary wave patterns to fill an orbit was only a preliminary stage half way between the Bohr model and our present quantum-mechanical one.

Now at this concluding stage of our programme, we should leave our pupils looking forward, not equipped with final knowledge – such as a 'modern, correct, true model of the atom at last' – but keen to see how knowledge and understanding grow as experiments continue and theories change.

We should give pupils a glimpse of our present atom-model: a nuclear atom with a fuzzy distribution of electrons instead of sharp orbits – fuzzy in position but definite in energy levels. The locations (and motions) of the electrons are described by their 'matter waves'. These wave patterns – which we write as equations when they are too difficult to sketch – tell us the *probability* of finding an electron in a given region of the atom. They tell us the betting, never a certainty. Yet the betting is useful: it tells us definite energy levels; it explains chemical bonding by electrons; and it not only explains the known random laws of radioactivity but also predicts new nuclear particles.

This *is* disturbing new knowledge; and we should not be sorry to leave our pupils at this point – they will remember science as combining experiments with a continuing changing series of models in our thinking which we call theory.

Appendix

BOOKS ON THE NEWER PHYSICS

In dealing with new, recent physics – the physics-in-the-making of the last few decades – we can only give some notes on teaching in this *Guide*. Many teachers would like to read fuller accounts of such topics as we have chosen. Yet when they look at books on modern physics they are disappointed. There are up-to-date advanced texts for university teaching and professional use; and there are some popular accounts of the latest physics, written for laymen. Many a book that gives the careful exposition of modern physics that one would like to have as background for O-Level teaching seems to stop short at the state of physics fifty years ago, or at least treats later topics too briefly. With that need in mind, we suggest the following books which might be useful:

The new age in physics by Sir Harrie Massey (Paul Elek, 1960)

The nature of solids by Alan Holden (Columbia University Press, 1965)

The great experiment in physics by H. S. Lipson (Oliver and Boyd, 1968)

The character of physical law by Richard Feynman (B.B.C., 1965)

Mr Tompkins in paperback by George Gamow (Cambridge U.P., 1965)

Guide to science by Isaac Asimov, *Vol 1: The physical sciences* (Penguin, 1975)

EXPERIMENTS AND PROJECTS FOR REVISION

The section on Astronomy (Chapters 3, 4, and 5) has been planned for pupil reading rather than for direct teaching. We suggest that much of the teaching time which might otherwise have been devoted to astronomy should be spent by pupils preparing demonstrations to show to one another. Pairs of pupils might be allowed to select an experiment from each of three lists like those which follow. We offer these lists as specimens; teachers will doubtless want to expand them with their own choices and the suggestions of the pupils.

List A includes a number of simple but fundamental experiments which pupils will either have seen or will have performed in earlier years. The tasks can easily be matched to the abilities of the pupils involved.

List B makes a second selection of experiments of a somewhat more complex nature, while List C offers suggestions for a range of simple projects.

If eight double periods can be allotted to this work, they might be grouped thus:

Week 1. Pupils working on choices from List A.

Week 2. Pupils present their List A experiment to the class as a whole.

Weeks 3, 4. Pupils working on choices from List B.

Week 5. Pupils present their List B choice to the class.

Weeks 6, 7. Pupils working on choices from List C, or alternatively on a second choice from either List A or List B.

Week 8. Pupils present this last experiment to the class.

Since the astronomy reading will be undertaken during this time we would think it unwise to expect written accounts of these revision experiments. The reward is in the doing and in the presentation to the class, as well as in the watching and thinking.

List A

Colour mixing

Crystal growing and crystal models

Currents and p.d.s in series and parallel circuits

The simple telescope

Lenses and mirrors in a smoke box

Many magnetic fields with (concealed) magnets

Inertia demonstrations

Ray streaks with lenses leading to the telescope model

Galileo's 'Pin and Pendulum'

Weighing air

The Brownian Motion in smoke

The catapult field

The model motor, dynamo, relay, etc.

Newton's 'Guinea and Feather'

Where is the image in a plane mirror?

Selections from the energy circus

Flask model in astronomy

List B

The oil film measurement
Magnetizing a ring of steel and showing success
Transformer demonstrations
Measurements on a d.c. power line
Measurements on an a.c. power line
The same force is applied to trolleys with masses 1, 2, and 3
Newton's second prism (do the colours in the spectrum split a second time?)
The Radiation Circus
Electroplating
Timing free fall with a scaler
Input and output power of f.h.p. motor
Bernoulli experiments
Diffraction at a straight edge
Change of volume of petrol on vaporization
Boyle's Law
Ripple tank to show reflection and refraction

List C

Does hot air conduct electricity?
Model a wave using trolleys and springs
Speed of sound using C.R.O.
Half and full wave rectification using C.R.O.
Using a relay
Checking the shutter speeds of a camera
Soap films on spirals and other shapes
Audible range of the human ear
'Pearls in air' for parabolic path
Illusions with colour
The strength of a human hair (or sewing cotton, or nylon thread, etc.)
Musical notes and a C.R.O.
Depth of field of a camera lens
Energy store in a catapult
Temperatures in a flame
The performance of a pinhole camera as the hole gets smaller and smaller

Index